Your Practice Set
Applications and Interpretation for IBDP Mathematics
Book 1
(For Both SL and HL Students)

Stephen Lee   Michael Cheung   Balance Lee

SE Production Limited

Your Practice Set
Applications and Interpretation for IBDP Mathematics
Book 1

Authors: Stephen Lee, Michael Cheung and Balance Lee

Published by
SE Production Limited
Website: www.seprodstore.com
Email: info@seprodstore.com

First Published Feb 2020
Published and Printed in Hong Kong
ISBN: 978-988-74134-2-4

All rights reserved. No part of this publication may be reproduced in whole or in part of transmitted in any form or by any means, electronic or mechanical, including photocopying, recording, or stored in any information storage and retrieval system, without permission in writing from the publisher.

This publication has been developed independently from and is not endorsed by the International Baccalaureate Organization. International Baccalaureate, Baccalauréat International, Bachillerato Internacional and IB are registered trademarks owned by the International Baccalaureate Organization.

Book cover: Mr. M. H. Lee

Authors

Stephen Lee, BSc (HKU), MStat (HKU), PGDE (CUHK)

Mr. Stephen Lee received his Bachelor of Science in Mathematics and Statistics, and Master of Statistics from The University of Hong Kong. During his postgraduate studies at HKU, he was a teaching assistant in the Department of Statistics and Actuarial Science, The University of Hong Kong, where he conducted tutorial lessons for undergraduate students. Later on, he received the Postgraduate Diploma of Education in Mathematics at from the Chinese University of Hong Kong. He is currently a frontline teacher in an IB World School. Apart from local syllabus in Hong Kong, he has experience in teaching various levels in IBDP Mathematics. He is also an examiner of the International Baccalaureate Organization (IBO). Furthermore, he is also the chief author of the book series: **Your Personal Coach Series – HKDSE Mathematics (Compulsory Part) Conventional Questions and Multiple Choice Questions,** and **Your Practice Set – Analysis and Approaches for IBDP Mathematics**.

Michael Cheung, BBA and Mathematics (HKUST), MSc in Mathematics (Universite Paris-Dauphine, France)

Mr. Michael Cheung has a strong Mathematics background and has been teaching Mathematics for more than 10 years. He conducted tutorial classes in fluent English to international students from different international schools. He has been teaching Mathematics in an IB world school. As an IB examiner, he needs to help on marking the IB exam papers every year. Based on his experience, he is very familiar with IB syllabus and knows about different question styles in real exam.

Balance Lee, BSc (CUHK), MStat (HKU)

Mr. Balance Lee received his Bachelor of Science in Risk Management Science from the Chinese University of Hong Kong, as well as the Master of Statistics in the University of Hong Kong. He has more than 10 years of experience in teaching students from various curricula notably the IBDP and the A level Mathematics syllabuses, including group courses conducted in English. He is currently a tutor mainly for IBDP Mathematics, and at the same time an examiner from the International Baccalaureate Organization, and keeping updated on the syllabus change in Mathematics.

Foreword

People in this world have different views on academic success. Some people think that academic success is measured by scores on examinations, while some may think that it should be measured by the happiness in learning. From my point of view, I think academic success is that students can learn in an effective way and have enjoyment in the learning process. Students can find learning interesting and have motivation if the learning process is effective, and thus learning becomes enjoyable and the chance of getting good academic results will be greater.

In preparing this book, our team was guided by our experience and interest in teaching IBDP Mathematics. This book is designed to help students to have a good preparation in the brand new challenging two-year International Baccalaureate Diploma Program. This book helps students to review all important concepts in Applications and Interpretation, and help students to understand how to start to answer a question and get familiar with assessment-styled questions. No doubt, this book can help you achieving high exam scores in IBDP Mathematics. By going through this book, you will find that the questions can help you to answer the structured questions confidently.

To sum up, this book is not only to be a successful practice source, but also to serve as valuable resource for students of each area.

SE Production Team

The main page of **SE Production Limited**

https://www.seprodstore.com   OR

The Facebook page of **SE Production Limited**

The Instagram page of **SE Production Limited**

The Twitter page of **SE Production Limited**

Please refer to the resource page of this book

https://www.seprodstore.com/ibaibook1material   OR

for any updates on this book.

Please feel free to email us through info@seprodstore.com if you find any error or if you have any suggestion on our product.

Contents

Authors
Foreword
Updates
Contents
More Recommendations
Ways to Use This Book
GDC Skills

Chapter 1	Standard Form	1
Chapter 2	Approximation and Error	4
Chapter 3	Functions	8
Chapter 4	Quadratic Functions	19
Chapter 5	Exponential and Logarithmic Functions	34
Chapter 6	Systems of Equations	43
Chapter 7	Arithmetic Sequences	53
Chapter 8	Geometric Sequences	62
Chapter 9	Financial Mathematics	72
Chapter 10	Coordinate Geometry	90
Chapter 11	Voronoi Diagrams	109
Chapter 12	Trigonometry	130
Chapter 13	2-D Trigonometry	144
Chapter 14	Areas and Volumes	170
Chapter 15	Differentiation	185
Chapter 16	Integration and Trapezoidal Rule	205
Chapter 17	Statistics	218
Chapter 18	Probability	243
Chapter 19	Discrete Probability Distributions	264
Chapter 20	Binomial Distribution	282
Chapter 21	Normal Distribution	292
Chapter 22	Bivariate Analysis	305
Chapter 23	Statistical Tests	331
	Answers	356

More Recommendations

Your Practice Set – Analysis and Approaches for IBDP Mathematics

- Common and compulsory topics for both MAA SL and MAA HL students
- 100 example questions + 400 intensive exercise questions in total
- 375 short questions + 125 structured long questions in total
- Special GDC skills included
- Holistic exploration on assessment styled questions
- QR Codes for online solution

Ways to Use This Book

SUMMARY POINTs	Checklist of the concepts of a particular topic for students
Paper 1 Questions	Short questions, usually 4 to 8 marks each
Paper 2 Questions	Structured questions, usually 12 to 20 marks each
[2]	Number of marks for a question
M1	A mark is assigned when the corresponding method is clearly shown
(M1)	A mark is assigned when the corresponding method is not clearly shown but is shown in the following correct working
A1	A mark is assigned when the correct answer is clearly shown
(A1)	A mark is assigned when the correct answer is not clearly shown but is shown in the following correct working
R1	A mark is assigned when the reasoning statement is clearly shown
N1	A mark is assigned when the correct answer is clearly shown, given that there is no working at all
AG	No mark is assigned as the final step (usually would be answer) is already given from the question

GDC Skills

Some implicit skills of TI-84 Plus CE that you might not heard before

Scenario 1: Solving $f(x) = g(x)$ in Functions

Step 1: Set $f(x) - g(x) = 0$
Step 2: Input $Y_1 = f(x) - g(x)$ in the graph function
Step 3: Set the screen size from window

> ✓ $x\min$ and $x\max$: You can refer to the domain given in the question
> ✓ $y\min$ and $y\max$: You can set $y\min = -1$ and $y\max = 1$ if you wish to find the x-intercept only

Scenario 2: Finding the number of years, n, when $f(x) = g(x)$ is in the exponent of an exponential model, in Arithmetic Sequences / Geometric Sequences / Logarithmic Functions

Step 1: Set the right-hand-side of the expression to be zero
Step 2: Input $Y_1 =$ the left-hand-side of the expression in the graph function
Step 3: Set the screen size from window

> ✓ $x\min$: You can set $x\min = 0$ as n represents the number of years which must be a positive integer

Scenario 3: Finding the x-intercept from the window

> ✓ Assume that the domain is $0 \leq x \leq 100$, and it is clearly shown that the curve cuts the x-axis once only on the left part of the screen
> ✓ You can set the left bound and the right bound to be 0 and 50 respectively to find the x-intercept efficiently, as 50 is the midpoint of the x-axis

Scenario 4: Finding an unknown quantity from the TVM Solver

N = 5
I% = 6
PV = −24000
PMT = 0
FV = ?
P/Y = 1
C/Y = 1
PMT : END

→

N = 5
I% = 6
PV = −24000
PMT = 0
FV = 0
P/Y = 1
C/Y = 1
PMT : END

- ✓ You can set the unknown quantity to be zero in order to execute the program. In the above example, the future value of a compound interest problem is going to be found. You can set FV to be zero and then choose tvm_FV to calculate the future value.

Scenario 5: Finding an area under a curve and above the x-axis

- ✓ Apart from using the function MATH 9, you can sketch the curve and use the function 2nd trace 7, and then set the lower limits and the upper limits.

Scenario 6: Finding probabilities in a Binomial distribution, in the form $P(X < \text{or} > \text{or} \geq c)$

- ✓ You need to change the probability to the form $P(X \leq C)$, and then use the function 2nd vars B to choose binomcdf.

Chapter

Standard Form

SUMMARY POINTs

✓ Standard Form: A number in the form $(\pm)a \times 10^k$, where $1 \leq a < 10$ and k is an integer

Solutions of Chapter 1

Your Practice Set – Applications and Interpretation for IBDP Mathematics

Paper 1 – Express Quantities in Standard Form

Example

A rectangle is 3250 cm long and 2720 cm wide.

(a) Find the perimeter of the rectangle, giving your answer in the form $a \times 10^k$, where $1 \leq a < 10$ and $k \in \mathbb{Z}$.

[2]

(b) Find the area of the rectangle, giving your answer in the form $a \times 10^k$, where $1 \leq a < 10$ and $k \in \mathbb{Z}$.

[2]

Solution

(a) The required perimeter
$= 2(3250 + 2750)$ (M1) for correct formula
$= 12000$
$= 1.2 \times 10^4$ cm   A1   N2

[2]

(b) The required area
$= 3250 \times 2750$ (M1) for correct formula
$= 8937500$
$= 8.9375 \times 10^6$ cm²   A1   N2

[2]

Exercise 1

1. For this question, give all the answers correct to 3 significant figures.

   The diameter of a circle is 1730 cm.

   (a) Find the circumference of the circle, giving your answer in the form $a \times 10^k$, where $1 \leq a < 10$ and $k \in \mathbb{Z}$.

   [2]

   (b) Find the area of the circle, giving your answer in the form $a \times 10^k$, where $1 \leq a < 10$ and $k \in \mathbb{Z}$.

   [2]

2. The base length and the height of a right-angled triangle are 3348 cm and 14880 cm respectively.

   (a) Find the length of the hypotenuse of the triangle, giving your answer in the form $a \times 10^k$, where $1 \leq a < 10$ and $k \in \mathbb{Z}$.

   [2]

   (b) Find the area of the triangle, giving your answer in the form $a \times 10^k$, where $1 \leq a < 10$ and $k \in \mathbb{Z}$.

   [2]

3. The base length and the area of a rectangle are 5476 cm and 22489932 cm² respectively.

   (a) Find the height of the rectangle, giving your answer in the form $a \times 10^k$, where $1 \leq a < 10$ and $k \in \mathbb{Z}$.

   [2]

   (b) Find the length of the diagonal of the rectangle, giving your answer in the form $a \times 10^k$, where $1 \leq a < 10$ and $k \in \mathbb{Z}$.

   [2]

4. The height and the area of a right-angled triangle are 8283 cm and 331320000 cm² respectively.

   (a) Find the base length of the triangle, giving your answer in the form $a \times 10^k$, where $1 \leq a < 10$ and $k \in \mathbb{Z}$.

   [2]

   (b) Find the length of the hypotenuse of the triangle, giving your answer in the form $a \times 10^k$, where $1 \leq a < 10$ and $k \in \mathbb{Z}$.

   [2]

Your Practice Set – Applications and Interpretation for IBDP Mathematics

Chapter

Approximation and Error

SUMMARY POINTs

✓ Summary of rounding methods:

2.71828	Correct to 3 significant figures	Correct to 3 decimal places
Round off	2.7**2**	2.71**8**

✓ Consider a quantity measured as Q and correct to the nearest unit d:

$\frac{1}{2}d$: Maximum absolute error

$Q - \frac{1}{2}d \leq A < Q + \frac{1}{2}d$: Range of the actual value A

$Q - \frac{1}{2}d$: Lower bound (Least possible value) of A

$Q + \frac{1}{2}d$: Upper bound of A

$\dfrac{\text{Maximum absolute error}}{Q} \times 100\%$: Percentage error

Solutions of Chapter 2

2 Paper 1 – Rounding and Percentage Error

Example

$$A = \frac{(2\sin(z))(\sqrt{x+17})}{64xy^2}, \text{ where } x = 10, y = 0.5 \text{ and } z = 60°.$$

(a) Calculate the **exact** value of A. [1]

(b) Give your answer to A correct to two significant figures. [1]

(c) Write down an inequality representing the error interval of this estimate. [2]

Casey estimates the value of A to be 0.055.

(d) Calculate the percentage error in Casey's estimate. [2]

Solution

(a) 0.05625   A1   N1 [1]

(b) 0.056   A1   N1 [1]

(c) $0.0555 \leq A < 0.0565$ A2   N2 [2]

(d) The percentage error
$$= \left|\frac{0.055 - 0.05625}{0.05625}\right| \times 100\%$$ (A1) for correct substitution
$$= 2.222222222\%$$
$$= 2.22\%$$ A1   N2 [2]

Your Practice Set – Applications and Interpretation for IBDP Mathematics

Exercise 2

1. $B = \dfrac{x\sqrt{y}}{\cos(90° - z)}$, where $x = 1.125$, $y = 1.5625$ and $z = 30°$.

   (a) Calculate the **exact** value of B.
   [1]

   (b) Give your answer to B correct to three significant figures.
   [1]

   (c) Write down an inequality representing the error interval of this estimate.
   [2]

   Julie estimates the value of B to be 2.84.

   (d) Calculate the percentage error in Julie's estimate.
   [2]

2. The lengths of the four sides of a quadrilateral are 5.278 cm, 4.812 cm, 4.118 cm and 3.756 cm respectively.

   (a) Calculate the **exact** perimeter of the quadrilateral.
   [2]

   The lengths of all four sides are estimated by rounding off, correct to 1 decimal place.

   (b) Write down the upper bound and the lower bound of the error interval of the estimate of the longest side.
   [2]

   (c) Calculate the percentage error in the estimate of the perimeter.
   [2]

3. The dimensions of a rectangular snack box are 15.75 cm, 8.95 cm and 7.15 cm.

   (a) Calculate the **exact** volume of the box.
   [2]

   The lengths of all sides of the box are estimated by rounding off, correct to the nearest cm.

   (b) Write down the upper bound and the lower bound of the error interval of the estimate of the shortest side.
   [2]

   (c) Calculate the percentage error in the estimate of the volume.
   [2]

4. For a farm in the shape of a right-angled triangle, the lengths of the hypotenuse and the shortest side are 39.063125 km and 10.937675 km respectively.

    (a) Calculate L, the **exact** length of the remaining side.

    [2]

    The lengths of all sides of the farm are estimated by rounding off, correct to the nearest 0.01 km.

    (b) Write down an inequality representing the error interval of the estimate of the remaining side.

    [2]

    (c) Calculate the percentage error in the estimate of the area.

    [2]

Chapter 3

Functions

SUMMARY POINTs

- The function $y = f(x)$:
    1. $f(a)$: Functional value when $x = a$
    2. Domain: Set of values of x
    3. Range: Set of values of y

- Steps of finding the inverse function $y = f^{-1}(x)$ of $f(x)$:
    1. Start from stating y in terms of x
    2. Interchange x and y
    3. Make y the subject in terms of x

- Properties of rational function $y = \dfrac{ax+b}{cx+d}$:
    1. $y = \dfrac{1}{x}$: Reciprocal function
    2. $y = \dfrac{a}{c}$: Horizontal asymptote
    3. $x = -\dfrac{d}{c}$: Vertical asymptote

Solutions of Chapter 3

3 Paper 1 – Sketching the Inverse Graph

Example

The diagram below shows the graph of a function f, for $0 \leq x \leq 4$.

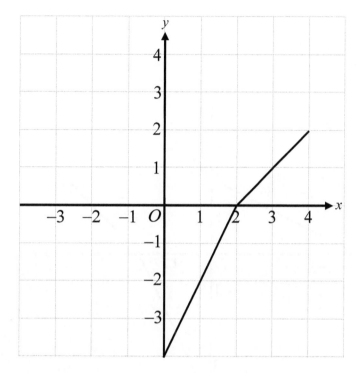

(a) Write down the value of

   (i) $f(1)$;

   (ii) $f^{-1}(2)$.

   [3]

(b) On the same diagram, sketch the graph of f^{-1}.

   [3]

The volume and the temperature of a particle can be modelled by the above function f, where x and $f(x)$ are the volume and the temperature of the particle respectively.

(c) Interpret the meaning of $f(4) = 2$.

   [1]

Your Practice Set – Applications and Interpretation for IBDP Mathematics

Solution

(a) (i) $f(1) = -2$ A1   N1

  (ii) $f^{-1}(2) = 4$ A2   N2

[3]

(b) For any two correct points from $(-4, 0)$, $(0, 2)$
   and $(2, 4)$ M1
   For correct graph   A2   N3

[3]

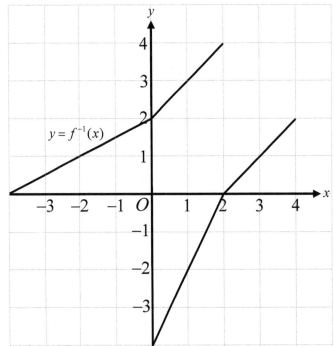

(c) When the volume of the particle is 4, its
   temperature is 2.   A1   N1

[1]

Exercise 3

1. The diagram below shows the graph of a function f, for $-4 \leq x \leq 3$.

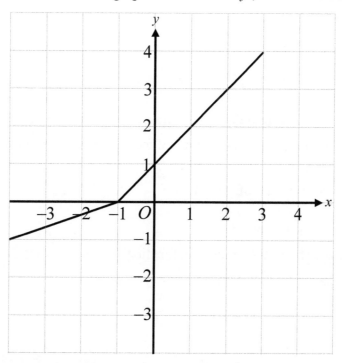

   (a) Write down the value of

   (i) $f(2)$;

   (ii) $f^{-1}(-1)$.

   [3]

   (b) On the same diagram, sketch the graph of f^{-1}.

   [3]

2. The diagram below shows the graph of a function f, for $-4 \leq x \leq 4$.

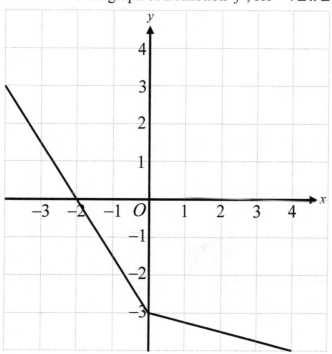

(a) Write down the value of

　　(i)　　$f(-4)$;

　　(ii)　　$f^{-1}(-4)$.

[3]

(b) On the same diagram, sketch the graph of f^{-1}.

[3]

The displacement and the velocity of a particle can be modelled by the above function f, where x and $f(x)$ are the displacement and the velocity of the particle respectively.

(c) Interpret the meaning of $f(-2) = 0$.

[1]

3. The diagram below shows the graph of a function f, for $-3 \le x \le 3$.

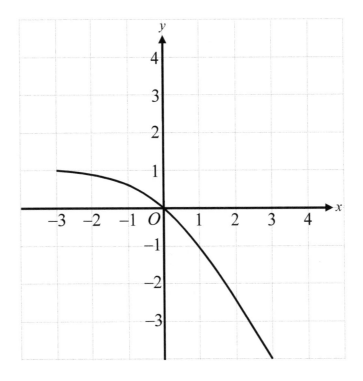

(a) On the same diagram, sketch the graph of f^{-1}.

[3]

(b) Write down the value of $f(1)$

[1]

(c) Write down the value of $f^{-1}(0)$

[1]

The range of f is $\{y : b \le y \le 1\}$.

(d) Write down the value of b.

[1]

4. The diagram below shows the graph of a function f, for $-5 \leq x \leq 4$.

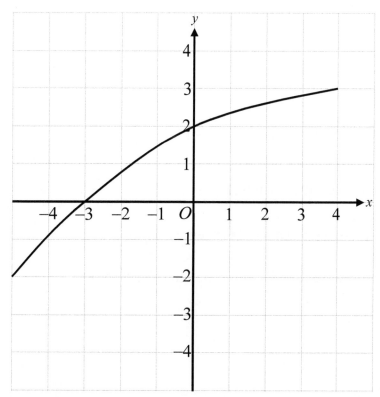

(a) On the same diagram, sketch the graph of f^{-1}.

[3]

(b) Write down the value of $f^{-1}(-1)$

[1]

The range of f is $\{y : -2 \leq y \leq c\}$.

(c) Write down the value of c.

[1]

The displacement and the velocity of a car can be modelled by the above function f, where x and $f(x)$ are the displacement and the velocity of the car respectively.

(d) Interpret the meaning of $f^{-1}(2) = 0$.

[1]

Your Practice Set – Applications and Interpretation for IBDP Mathematics

 Paper 1 – Rational Functions

Example

Consider the graph of the function $f(x) = 1 - \dfrac{4}{x}$, $x \neq 0$.

(a) Find the value of x such that $f(x) = 0$.
[2]

(b) Write down the equation of the vertical asymptote.
[2]

(c) Write down the equation of the horizontal asymptote.
[2]

Solution

(a) $f(x) = 0$

$1 - \dfrac{4}{x} = 0$ (M1) for setting equation

$1 = \dfrac{4}{x}$

$x = 4$ A1   N2
[2]

(b) $x = 0$ A2   N2
[2]

(c) $y = 1$ A2   N2
[2]

Exercise 4

1. Consider the graph of the function $f(x) = \dfrac{2}{x+3} - 5$, $x \neq -3$.

   (a) Find the value of x such that $f(x) = 0$.
   [2]

   (b) Write down the equation of the vertical asymptote.
   [2]

   (c) Write down the equation of the horizontal asymptote.
   [2]

2. Consider the graph of the function $f(x) = 3 - \dfrac{6}{1-x}$, $x \neq 1$.

    (a) Find the y-intercept of the graph.

        [2]

    (b) Write down the equation of the vertical asymptote.

        [2]

    (c) Write down the equation of the horizontal asymptote.

        [2]

3. Consider the graph of the function $f(x) = \dfrac{x+2}{x-5}$, $x \neq 5$.

    (a) Write down the equation of the vertical asymptote.

        [2]

    (b) Write down the equation of the horizontal asymptote.

        [2]

    (c) Write down the domain of $f(x)$.

        [1]

    (d) Let $g(x) = x$. Write down the range of values of x such that $f(x) > g(x)$.

        [2]

4. Consider the graph of the function $f(x) = \dfrac{7x-1}{2x-5}$, $x \neq \dfrac{5}{2}$.

    (a) Write down the equation of the vertical asymptote.

        [2]

    (b) Write down the equation of the horizontal asymptote.

        [2]

    (c) Write down the range of $f(x)$.

        [1]

    (d) Let $g(x) = x - 2$. Write down the range of values of x such that $f(x) \leq g(x)$.

        [2]

Your Practice Set – Applications and Interpretation for IBDP Mathematics

Paper 1 – Linear Functions

Example

The relationship between F, the temperature in degrees Fahrenheit, and C, the temperature in degrees Celsius, is given by $F = mC + 32$.

(a) Interpret what m represents.

[1]

It is given that the temperature in 25 degrees Celsius is equivalent to the temperature in 77 degrees Fahrenheit.

(b) Find the value of m.

[2]

(c) Find the value of 122 degrees Fahrenheit in degrees Celsius.

[2]

Solution

(a) m represents the rate of change of degrees Fahrenheit per 1 degree Celsius increase.   A1   N1

[1]

(b) $77 = m(25) + 32$ (M1) for substitution

$45 = 25m$

$m = \dfrac{9}{5}$ A1   N2

[2]

(c) $122 = \dfrac{9}{5}C + 32$ (M1) for substitution

$90 = \dfrac{9}{5}C$

$C = 50$

Therefore, the temperature is 50 degrees Celsius.   A1   N2

[2]

Exercise 5

1. The relationship between B, the boiling point of water in degrees Celsius, and x, the vertical height above the sea level in metres, is given by $B = 100 + mx$.

   (a) Interpret what m represents.
   [1]

   It is given that the boiling point of water is 64 degrees Celsius at a height of 10 km.

   (b) Find the value of m.
   [2]

   (c) If water boils at the top of the Alphubel in Switzerland at 84 degrees Celsius, find the height of the top of the Alphubel above the sea level.
   [2]

2. The number of hotels in a city, N, can be modelled by $N = at + b$, where t is the number of years after 2007.

   (a) Interpret what b represents.
   [1]

   It is given that there are 143 hotels in 2007 and 193 hotels in 2012.

   (b) Find the values of a and b.
   [3]

   (c) Find the number of hotels in 2015.
   [2]

3. The daily salary of a worker is modelled by $S = at + b$, where $\$S$ is the salary of the worker and t represents his working time in hours.

   (a) Interpret what a represents.
   [1]

   (b) Interpret what b represents.
   [1]

   Stephen earns $600 after he has worked for 8 hours on a particular day, and the fixed daily salary is $200.

   (c) Find the values of a and b.
   [3]

   (d) Find the daily salary if a worker only works for 30 minutes per day.
   [2]

Your Practice Set – Applications and Interpretation for IBDP Mathematics

4. An aluminium lamina of area A mm^2 is heated. Its temperature $T°$C can be modelled by $A = pT + q$.

   (a) Interpret what p represents.

   [1]

   (b) Interpret what q represents.

   [1]

   It is given that the areas of the aluminium lamina are 5 mm^2 and 8 mm^2 at 0°C and 60°C respectively.

   (c) Find the values of p and q.

   [3]

   (d) Find the difference of the areas of the aluminium lamina after a drop of 140°C.

   [2]

Chapter

Quadratic Functions

SUMMARY POINTs

✓ General form $y = ax^2 + bx + c$, where $a \neq 0$:

$a > 0$	The graph opens upward
$a < 0$	The graph opens downward
c	y-intercept
$h = -\dfrac{b}{2a}$	x-coordinate of the vertex
$k = ah^2 + bh + c$	y-coordinate of the vertex
	Extreme value of y
$x = h$	Equation of the axis of symmetry

✓ Other forms:
1. $y = a(x-h)^2 + k$: Vertex form
2. $y = a(x-p)(x-q)$: Intercept form with x-intercepts p and q

Solutions of Chapter 4

6 Paper 1 – x-intercepts and Coordinates of Vertex

Example

Let $f(x) = 3x^2 - 12x - 15$. Part of the graph of f is shown below.

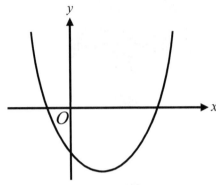

(a) Find the x-intercepts of the graph.

[4]

(b) (i) Write down the equation of the axis of symmetry.

(ii) Find the y-coordinate of the vertex.

[3]

Solution

(a) $f(x) = 0$ (M1) for function equals to 0

$3x^2 - 12x - 15 = 0$

$3(x+1)(x-5) = 0$ A1

$x = -1$ or $x = 5$

Thus, the x-intercepts are -1 and 5.   A2   N4

[4]

(b) (i) $x = 2$ A1   N1

(ii) The y-coordinate of the vertex

$= 3(2)^2 - 12(2) - 15$ (M1) for substitution

$= -27$ A1   N2

[3]

Exercise 6

1. Let $f(x) = x^2 - 6x + 8$. Part of the graph of f is shown below.

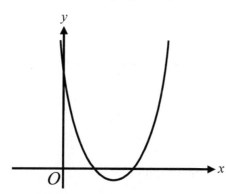

   (a) Find the x-intercepts of the graph.

   [4]

   (b) (i) Write down the equation of the axis of symmetry.

   (ii) Find the y-coordinate of the vertex.

   [3]

2. Let $f(x) = x^2 - 11x + 10$. Part of the graph of f is shown below.

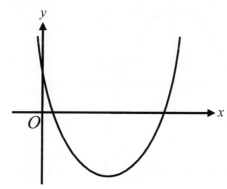

   (a) Find the x-intercepts of the graph.

   [4]

   (b) (i) Write down the equation of the axis of symmetry.

   (ii) Find the y-coordinate of the vertex.

   [3]

3. Let $f(x) = -2x^2 - 14x$. Part of the graph of f is shown below.

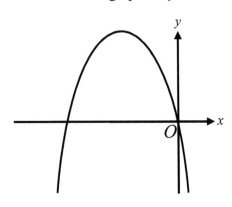

(a) Find the x-intercepts of the graph.

[4]

(b) (i) Write down the equation of the axis of symmetry.

(ii) Find the y-coordinate of the vertex.

[3]

4. Let $f(x) = 13.5 - 1.5x^2$. Part of the graph of f is shown below.

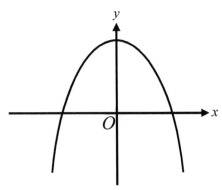

(a) Find the x-intercepts of the graph.

[4]

(b) (i) Write down the equation of the axis of symmetry.

(ii) Find the y-coordinate of the vertex.

[3]

Paper 1 – Factorized Form

Example

Let $f(x) = x^2 + 2x - 8$.

(a) Factorize $f(x)$.

[2]

(b) Write down the x-intercepts of the graph of f.

[2]

$f(x)$ reaches its minimum value at $y = -9$.

(c) Write down the range of f.

[1]

Solution

(a) $f(x) = (x-2)(x+4)$ A2   N2

[2]

(b) $x = 2$ and $x = -4$ A2   N2

[2]

(c) $\{y : y \geq -9\}$ A1   N1

[1]

Exercise 7

1. Let $f(x) = x^2 - 2x - 35$.

    (a) Factorize $f(x)$.

    [2]

    (b) Write down the x-intercepts of the graph of f.

    [2]

    $f(x)$ reaches its extreme value at $y = -36$.

    (c) Write down the range of f.

    [1]

Your Practice Set – Applications and Interpretation for IBDP Mathematics

2. Let $f(x) = -2x^2 - 14x - 12$.

   (a) Factorize $f(x)$.

   [2]

   (b) Write down the x-intercepts of the graph of f.

   [2]

   $f(x)$ reaches its extreme value at $y = 12.5$.

   (c) Write down the range of f.

   [1]

3. Let $f(x) = a(x-p)(x-q)$.

   The graph of $f(x)$ passes through the points $(5, 0)$, $(10, -7.5)$ and $(11, 0)$.

   (a) Write down the value of p and of q.

   [2]

   (b) Find the value of a.

   [3]

   The y-coordinate of the vertex of $f(x)$ is -13.5.

   (c) Write down the range of f.

   [1]

4. Let $f(x) = a(p-x)(x-q)$.

   The graph of $f(x)$ passes through the origin, $(15, 30)$ and $(18, 0)$.

   (a) Write down the value of p and of q.

   [2]

   (b) Find the value of a.

   [3]

   The y-coordinate of the vertex of $f(x)$ is 54.

   (c) Write down the range of f.

   [1]

8 Paper 1 – Symmetric Properties

Example

The graph of a quadratic function $y = f(x)$ is given below.

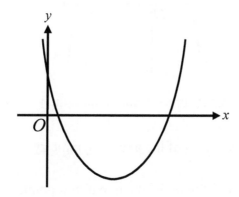

It is given that the x-intercepts are 2 and 18, and the y-coordinate of the vertex is -25.

(a) Write down the equation of the axis of symmetry.

[2]

(b) Write down the coordinates of the minimum point.

[1]

(c) Write down the range of $f(x)$.

[2]

Solution

(a) $x = 10$ A2   N2

[2]

(b) $(10, -25)$ A1   N1

[1]

(c) $\{y : y \geq -25\}$ A2   N2

[2]

Your Practice Set – Applications and Interpretation for IBDP Mathematics

Exercise 8

1. The graph of a quadratic function $y = f(x)$ is given below.

   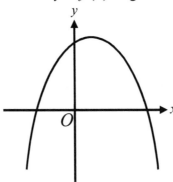

   It is given that the x-intercepts are -4 and 8, and the y-coordinate of the vertex is 17.

   (a) Write down the equation of the axis of symmetry.

   [2]

   (b) Write down the coordinates of the maximum point.

   [1]

   (c) Write down the range of $f(x)$.

   [2]

2. The graph of a quadratic function $y = f(x)$ is given below.

   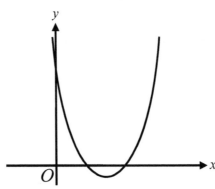

   It is given that one of the x-intercepts is 10 and the equation of the axis of symmetry is $x = 7$.

   (a) Write down the other x-intercept.

   [2]

   It is given that $f(7) = -2$.

   (b) Write down the coordinates of the vertex.

   [1]

   (c) Write down the range of $f(x)$.

   [2]

3. The graph of a quadratic function $y = ax(x+r)$ is given below.

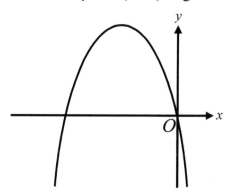

It is given that the equation of the axis of symmetry is $x = -5$ and the range of $f(x)$ is $\{y : y \leq 12.5\}$.

(a) Write down the coordinates of the vertex.

[1]

(b) Find the values of a and r.

[5]

4. The graph of a quadratic function $y = f(x)$ is given below.

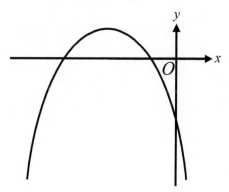

It is given that the equation of the axis of symmetry is $x = -2.5$ and $f(0) = f(s) = -4$.

(a) Write down the value of s.

[1]

It is given that $y = a(x+r)(x+1)$.

(b) Find the values of a and r.

[4]

(c) It is given that $f(2) = t$. Write down the value of $f(-7)$ in terms of t.

[1]

Your Practice Set – Applications and Interpretation for IBDP Mathematics

Paper 1 – Inequalities Involving Quadratic Functions

Example

A quadratic function is given by $y = (x-5)^2 + 10$.

(a) Find the y-intercept of the function.

[2]

The value of y is restricted such that $y \leq 19$. In this situation, the values of x lies between p and 8, where $p < 8$.

(b) Find the value of p.

[2]

(c) Write down the value of x when y attains its minimum.

[1]

Solution

(a) The y-intercept
$= (0-5)^2 + 10$ (M1) for valid approach
$= 35$ A1   N2

[2]

(b) $(x-5)^2 + 10 \leq 19$
$(x-5)^2 - 9 \leq 0$ (M1) for setting inequality

By considering the graph of $y = (x-5)^2 - 9$,
$2 \leq x \leq 8$.
$\therefore p = 2$ A1   N2

[2]

(c) 5   A1   N1

[1]

Exercise 9

1. A quadratic function is given by $y = -(x-100)^2 + 80$.

   (a) Find the y-intercept of the function.
   [2]

   The value of y is restricted such that $y \geq 16$. In this situation, the values of x lies between 92 and p, where $p > 92$.

   (b) Find the value of p.
   [2]

   (c) Write down the value of x when y attains its maximum.
   [1]

2. A quadratic function is given by $y = x^2 - 40x + 5400$.

   (a) Write down the y-intercept of the function.
   [1]

   The value of y is restricted such that $5064 \leq y \leq 5400$. In this situation, the values of x lies between p and q, where $20 < p < q$.

   (b) Find the values of p and q.
   [3]

   (c) Write down the coordinates of the vertex.
   [2]

3. The cost C of producing x books by SE Production can be modelled by $C = 0.5(x-60)^2 + 40$.

   (a) Find the cost of producing 200 books.
   [2]

   To limit the total cost, the upper limit for the cost is set to be 240.

   (b) Find the range of the number of books.
   [2]

   (c) Write down the minimum cost.
   [1]

   (d) Write down the number of books produced when the cost is at its minimum.
   [1]

Your Practice Set – Applications and Interpretation for IBDP Mathematics

4. The average profit P of selling x buckets of fruits can be modelled by $P = -0.25(x-20)^2 + 21$.

    (a) Find the average profit of selling 22 buckets of fruits.

    [2]

    A target is set such that the average profit should be at least 17.

    (b) Find the range of the number of buckets of fruits.

    [2]

    Due to limitation, the number of buckets of fruits to be sold is at most 18.

    (c) Find the maximum average profit for this limitation.

    [2]

Paper 1 – Applications of Quadratic Functions

Example

The length of a square farm is $(x+2)$ m. In two of the four corners a square of length 2 m is for storage use. The rest of the farm is for animals. Let A m^2 be the area of the region for animals.

(a) Find the expression of A in terms of x.

[2]

It is given that $A = 174.25$.

(b) Find the value of x.

[3]

(c) Write down the perimeter of the region for animals.

[1]

Solution

(a) A
$= (x+2)^2 - 2(2)^2$ (M1) for valid approach
$= x^2 + 4x + 4 - 8$
$= x^2 + 4x - 4$ A1   N2

[2]

(b) $A = 174.25$
$x^2 + 4x - 4 = 174.25$ (M1) for setting equation
$x^2 + 4x - 178.25 = 0$
$4x^2 + 16x - 713 = 0$ (A1) for correct equation
$(2x + 31)(2x - 23) = 0$
$2x + 31 = 0$ or $2x - 23 = 0$
$x = -15.5$ (*Rejected*) or $x = 11.5$ A1   N3

[3]

(c) 54 m A1   N1

[1]

Your Practice Set – Applications and Interpretation for IBDP Mathematics

Exercise 10

1. The length and the width of a rectangular board are 30 cm and 20 cm respectively. The four corners in a square shape of length x cm is removed. The remaining part of the rectangular board forms a rectangular box of height x cm as shown in the diagram below. Let A cm^2 be the area of the base of the box.

    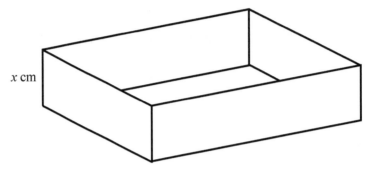

    (a) Find the expression of A in terms of x.
    [2]

    It is given that $A = 299$.

    (b) Find the value of x.
    [3]

    (c) Write down the exact value of the volume of the box.
    [1]

2. In a right-angled triangle, the lengths of the three sides are $(x-18)$ cm, $(x-1)$ cm and x cm respectively. Let A cm^2 be the area of the triangle.

    (a) Write down and simplify a quadratic equation in x.
    [2]

    (b) Hence, find the value of x.
    [2]

    The triangle is then planned to be painted, and the total painting cost is $1680.

    (c) Find the painting cost per cm^2.
    [2]

3. In a coffee shop, the amount of coffee powder, Q, in kilograms, sold in a week is given by $Q = 75 - 15r$, where $\$r$ is the cost of 1 kg coffee bean.

   (a) Find the maximum cost of 1 kg coffee bean if at least 7.5 kg of coffee powder can be sold in a week.
   [2]

   The coffee shop can earn $\$(r+2)$ for each kilogram of coffee powder sold. Let P be the weekly profit.

   (b) Find the expression of P in terms of r.
   [2]

   (c) Find the exact value of the maximum weekly profit.
   [2]

4. A ball is kicked from the top of a vertical cliff onto a horizontal grass ground. The path of the ball can be modelled by the quadratic curve $y = -0.5x^2 + 2x + 10$, where x m and y m are the horizontal distance from the cliff and the vertical distance above the ground respectively, as shown in the diagram below.

   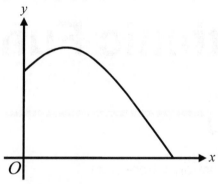

   (a) Find the maximum height of the trajectory of the ball.
   [2]

   (b) Write down the horizontal distance of the ball from the cliff when the ball is at the same vertical level when the ball is first kicked.
   [1]

   (c) Find the horizontal distance from the cliff to the position at which the ball hits the grass ground.
   [3]

Your Practice Set – Applications and Interpretation for IBDP Mathematics

Chapter 5

Exponential and Logarithmic Functions

SUMMARY POINTs

- ✓ $y = a^x$: Exponential function, where $a \neq 1$

- ✓ $y = \log_a x$: Logarithmic function, where $a > 0$

- ✓ $y = \log x = \log_{10} x$: Common Logarithmic function

- ✓ Properties of the graphs of $y = a^x$:

$a > 1$	$0 < a < 1$
y-intercept $= 1$	
y increases as x increases	y decreases as x increases
y tends to zero as x tends to negative infinity	y tends to zero as x tends to positive infinity
Horizontal asymptote: $y = 0$	

Solutions of Chapter 5

Paper 1 – Exponential Functions

Example

The function $f(x)$ is defined as $f(x) = 13 + a \times 2^{bx}$, where $b < 0$. The graph of $f(x)$ intersects the y-axis at $(0, 19)$.

(a) Find the value of a.

[2]

The graph of $f(x)$ passes through $(1, 13.75)$.

(b) Find the value of b.

[2]

(c) Write down the equation of the horizontal asymptote of the graph of $f(x)$.

[2]

Solution

(a) $19 = 13 + a \times 2^{b(0)}$ (M1) for substitution
       $19 = 13 + a$
       $a = 6$ A1   N2

[2]

(b) $13.75 = 13 + 6 \times 2^{b(1)}$ (M1) for substitution
       $0.75 - 6 \times 2^b = 0$
       By considering the graph of $y = 0.75 - 6 \times 2^b$,
       $b = -3$.   A1   N2

[2]

(c) $y = 13$ A2   N2

[2]

Your Practice Set – Applications and Interpretation for IBDP Mathematics

Exercise 11

1. The function $f(x)$ is defined as $f(x) = a - b^{-x}$, where $a, b > 0$. The graph of $f(x)$ intersects the y-axis at $(0, 9)$.

    (a) Find the value of a.

    [2]

    The graph of $f(x)$ passes through $\left(3, \dfrac{262}{27}\right)$.

    (b) Find the value of b.

    [2]

    (c) Write down the equation of the horizontal asymptote of the graph of $f(x)$.

    [2]

2. The function $g(x)$ is defined as $g(x) = p \times q^x + 7$, where $p, q > 0$. The graph of $g(x)$ intersects the y-axis at $(0, 11)$.

    (a) Find the value of p.

    [2]

    The graph of $g(x)$ passes through $(2, 8)$.

    (b) Find the value of q.

    [2]

    (c) Write down the range of $f(x)$.

    [2]

3. The function $g(x)$ is defined as $g(x) = 3 \times 2^{-px} + q$, where $p > 0$. The graph of $g(x)$ passes through $(0, 2)$ and $(-2, 47)$.

    (a) Find the values of p and q.

    [4]

    (b) Find the x-intercept of the graph of $g(x)$.

    [2]

    (c) Write down the number of solutions of $g(x) = -2$.

    [1]

4. The function $f(x)$ is defined as $f(x) = 2 \times a^x + b$, where $a, b > 0$. The graph of $f(x)$ passes through $(0, 6)$ and $(4, 166)$.

    (a) Find the values of a and b.

    [4]

    (b) Write down the change of the values of $f(x)$ if the values of the elements in domain increases.

    [1]

    (c) Write down the number of solutions of $f(x) = 10$.

    [1]

Your Practice Set – Applications and Interpretation for IBDP Mathematics

Paper 1 – Exponential Models

Example

The number of elephants P in a small jungle is modelled by $P(t) = 170 - 90e^{-\frac{1}{35}t}$, where t is the number of years after 2015.

(a) Write down the number of elephants in 2015.
[1]

(b) Find the number of elephants in 2019, giving the answer correct to the nearest integer.
[2]

(c) Find the number of years needed after 2015 when the number of elephants first exceeds 140.
[2]

(d) Write down the minimum value of the upper bound of the number of elephants.
[1]

Solution

(a) 80   A1   N1
[1]

(b) The number of elephants
$= 170 - 90e^{-\frac{1}{35}(4)}$ (M1) for substitution
$= 89.71972447$
$= 90$ A1   N2
[2]

(c) $170 - 90e^{-\frac{1}{35}t} > 140$ (M1) for valid approach
$30 - 90e^{-\frac{1}{35}t} > 0$

By considering the graph of $y = 30 - 90e^{-\frac{1}{35}t}$,
$t = 38.45143$.
Thus, the number of years needed is 38.5 years.   A1   N2
[2]

(d) 170   A1   N1
[1]

Exercise 12

1. The price P of a new car made in Germany is modelled by $P(t) = 2000 + 16000e^{-\frac{t}{8}}$, where t is the number of years after the car is purchased.

    (a) State the price of the car at the time it is purchased.
    [1]

    (b) Find the price of the car after 6 years.
    [2]

    (c) Find the time taken for the price of the car falls below $7000.
    [2]

    (d) Write down the maximum value of the lower bound of the price of the car in the long run.
    [1]

2. The number of bacteria N in an experiment box is modelled by $N(t) = 115 + 155e^{\frac{t}{25}}$, where t is the time in days from the start of the experiment.

    (a) Write down the number of bacteria at the beginning of the experiment.
    [1]

    (b) Find the increase in the number of bacteria in the first week, giving the answer correct to the nearest integer.
    [2]

    The maximum capacity of the experiment box is 1200 units of bacteria.

    (c) Find the number of days for the experiment box to reach its maximum capacity.
    [2]

3. The price EUR P of a computer system is modelled by $P(t) = 90 + A \times 0.7^t$, where A is a constant and t is the age of the computer measured in years. The initial price of the computer system is EUR 840.

    (a) Interpret the term $(90 + A)$ represents in this context.
    [1]

    (b) Find the value of A.
    [2]

    (c) Find the time when the price of the computer system drops to half of the initial price.
    [2]

    (d) Write down the maximum value of the lower bound of the price of the computer system in the long run.
    [1]

Your Practice Set – Applications and Interpretation for IBDP Mathematics

4. The amount of electric charge A stored in a battery of a tablet computer is modelled by $A(t) = p - q^t$, where p and q are positive constants, and t is the time in hours for the battery to be charged.

   (a) Interpret the term $(p-1)$ represents in this context.

   [1]

   The amount of electric charge at $t=0$ and $t=3$ are 4 units and 4.488 units respectively.

   (b) Find the values of p and q.

   [4]

   To use the tablet computer to watch an online concert, an electric charge of 4.3 units is needed.

   (c) Find the time required for the tablet computer to reach an electric charge of 4.3 units.

   [2]

Paper 1 – Logarithmic Models

Example

The magnitude of an earthquake can be modelled by the following function

$$M = 0.5 + \log_{10} P, \ P \geq 1$$

where M represents the magnitude of the earthquake and P represents the relative energy released.

(a) Find the magnitude of an earthquake if 2300 units of relative energy are released in the earthquake.

[2]

(b) Find the amount of relative energy released in an earthquake if the corresponding magnitude of the earthquake is 5.1.

[2]

Solution

(a) The required magnitude
$= 0.5 + \log_{10} 2300$ (M1) for correct formula
$= 3.861727836$
$= 3.86$ A1   N2

[2]

(b) $5.1 = 0.5 + \log_{10} P$ (M1) for substitution
$4.6 = \log_{10} P$
$\log_{10} P - 4.6 = 0$

By considering the graph of $y = \log_{10} P - 4.6$,
$P = 39810.71706$.
Thus, the amount of relative energy is 39800 units.   A1   N2

[2]

Your Practice Set – Applications and Interpretation for IBDP Mathematics

Exercise 13

1. A scientist defined the magnitude of an explosion by

   $$N = 3\log_{10}(2E), \quad N \geq 0.5$$

   where N represents the magnitude of the explosion and E represents the amount of energy released during the explosion.

   (a) Find the magnitude of an explosion if 5×10^3 units of energy are released during the explosion.

   [2]

   (b) Find the amount of energy released during the explosion if the corresponding magnitude of the explosion is 0.9.

   [2]

2. A researcher defined the sound intensity level in an explosion by

   $$Q = 120 + 9.9\log_{10} I, \quad I \geq 0$$

   where Q dB represents the sound intensity level and I Wm^{-2} represents the sound intensity.

   If the sound intensity level increases from 139.8 dB to 169.5 dB, find the ratio of the new sound intensity to the original sound intensity.

   [5]

3. Consider the function $f(x) = 2x + 3\log_{10}(x+2)$, where $x > -2$.

   (a) Find the y-intercept of the function $f(x)$.

   [2]

   (b) Write down the number of x-intercept(s) of the function $f(x)$.

   [1]

   (c) Solve the equation $f(x) = x^2$.

   [3]

4. Consider the function $f(x) = x + \log_{10}(x+10) + 2$, where $x > -10$. The graph of $f(x)$ intersects the x-axis and the y-axis at A and B respectively. Let O be the origin. Find the area of the triangle OAB.

   [6]

Chapter

Systems of Equations

SUMMARY POINTs

- ✓ $\begin{cases} ax + by = c \\ dx + ey = f \end{cases}$: 2×2 system

- ✓ $\begin{cases} ax + by + cz = d \\ ex + fy + gz = h \\ ix + jy + kz = l \end{cases}$: 3×3 system

- ✓ The above systems can be solved by PlySmlt2 in TI-84 Plus CE

Solutions of Chapter 6

14 Paper 1 – Mapping Diagrams

Example

The mapping diagram below represents the function $f(x) = ax + b$.

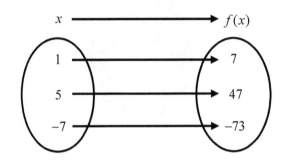

(a) Write down two equations in terms of a and b.

[2]

(b) Find the values of a and b.

[2]

(c) Find the value of c when $f(c) = c$.

[2]

Solution

(a) $a + b = 7$ A1   N1
$5a + b = 47$ A1   N1
(or $-7a + b = -73$)

[2]

(b) $a = 10$, $b = -3$ A2   N2

[2]

(c) $c = 10c - 3$ (M1) for substitution
$-9c = -3$
$c = \dfrac{1}{3}$ A1   N2

[2]

Exercise 14

1. The mapping diagram below represents the function $f(x) = ax^2 + bx$.

   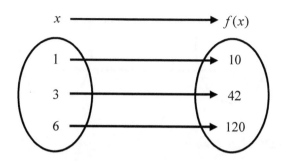

   (a) Write down two equations in terms of a and b. [2]

   (b) Find the values of a and b. [2]

   (c) Find the equation of the axis of symmetry of f. [2]

2. The mapping diagram below represents the function $f(x) = ax^3 + b$.

   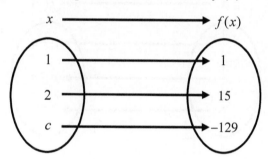

   (a) Find the values of a and b. [4]

   (b) Hence, find the value of c. [2]

3. The mapping diagram below represents the function $f(x) = \dfrac{a}{3-x}$.

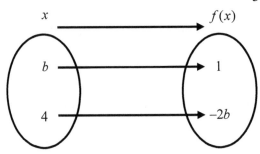

(a) Find the values of a and b.

[4]

(b) List the elements in the domain of f.

[1]

(c) Write down the equation of the horizontal asymptote.

[1]

4. The mapping diagram below represents the function $f(x) = \dfrac{1}{x^2}$.

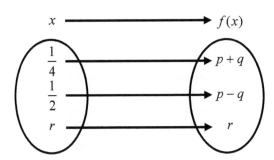

(a) Find the values of p, q and r.

[6]

(b) List the elements in the range of f.

[1]

(c) Write down the equation of the vertical asymptote.

[1]

Paper 1 – 2×2 Systems

Example

3000 citizens attended a carnival. Let x be the number of adults attending the carnival and y be the number of children attending the carnival.

(a) Write down an equation in x and y.

[1]

The cost of an adult ticket and a child ticket were set to be USD 18 and USD 8 respectively. The total cost of tickets sold in the carnival was USD 36000.

(b) Write down another equation in x and y.

[1]

(c) Write down the values of x and y.

[2]

(d) Find the total cost for a group of 4 adults and 7 children.

[2]

Solution

(a) $x + y = 3000$ A1   N1

[1]

(b) $18x + 8y = 36000$ A1   N1

[1]

(c) $x = 1200$, $y = 1800$ A2   N2

[2]

(d) The total cost
$= 18(4) + 8(7)$ (M1) for substitution
$= $ USD 128   A1   N2

[2]

Your Practice Set – Applications and Interpretation for IBDP Mathematics

Exercise 15

1. The Division 1 Football League in Japan has 30 matches for every participating team. The champion of the League last year is undefeated for all 30 matches. Let x and y be the number of wins and draws of the champion team respectively.

    (a) Write down an equation in x and y.

    [1]

    A team gets 3 points for a win and 1 point for a draw. The champion team got 82 points last year.

    (b) Write down another equation in x and y.

    [1]

    (c) Write down the values of x and y.

    [2]

    (d) Find the total points of an undefeated team if the number of wins and the number of draws are the same.

    [2]

2. The length, L cm, of a heated copper plate is given by $L = a + bT$, where T °C is the temperature measured. At 190 °C, the length of the copper plate is 8.25 cm.

    (a) Write down an equation in a and b.

    [1]

    At 220 °C, the length of the copper plate is 9.21 cm.

    (b) Write down another equation in a and b.

    [1]

    (c) Write down the values of a and b.

    [2]

    (d) Find the temperature of the copper plate when its length is 98.5 mm.

    [2]

3. The number of flats y can be modelled by the equation $y = pt + q$, where t is the number of years after 1 January, 2010.

   On 1 January 2012, the number of flats is 18000. On 1 January 2017, the number of flats is 22000.

   (a) Find the values of p and q.
   [4]

   (b) State the meaning of p in this context.
   [1]

   (c) State the meaning of q in this context.
   [1]

4. Laurent can buy 10 CDs and 7 DVDs for USD 76.5, and he can buy 8 CDs and 11 DVDs for USD 90.9.

   (a) Find the price of one CD and that of one DVD.
   [4]

   (b) Laurent wants to buy 7 CDs and 10 DVDs. Find the amount of change if he pays USD 100 at the counter.
   [2]

16 Paper 1 – 3×3 Systems

Example

The total profit $y of selling plastic boxes can be modelled by the equation $y = ax^2 + bx + c$, where x is the number of plastic boxes ordered from a factory, and a, b and c are real numbers. The total profits of selling 100 plastic boxes, 200 plastic boxes and 400 plastic boxes are $122, $424 and $1628 respectively.

(a) (i) Show that $10000a + 100b + c = 122$.

(ii) Show that $40000a + 200b + c = 424$.

(iii) Write down the third equation in a, b and c.

[3]

(b) Hence, find the values of a, b and c.

[4]

Solution

(a) (i) $122 = a(100)^2 + b(100) + c$ A1
$10000a + 100b + c = 122$ AG   N0

(ii) $424 = a(200)^2 + b(200) + c$ A1
$40000a + 200b + c = 424$ AG   N0

(iii) $160000a + 400b + c = 1628$ A1   N1

[3]

(b) $\begin{cases} 10000a + 100b + c = 122 \\ 40000a + 200b + c = 424 \\ 160000a + 400b + c = 1628 \end{cases}$ (M1) for valid approach

$a = \dfrac{1}{100}$, $b = \dfrac{1}{50}$ and $c = 20$ A3   N4

[4]

Exercise 16

1. The height h metres of a parachute t seconds after released from a helicopter can be modelled by $h = at^2 + bt + c$, and a, b and c are real numbers. The heights of a parachute at different times are shown in the following table:

t (s)	1	3	6
h (m)	998	982	928

   (a) (i) Show that $a + b + c = 998$.

       (ii) Show that $9a + 3b + c = 982$.

       (iii) Write down the third equation in a, b and c.

   [3]

   (b) Hence, find the values of a, b and c.

   [4]

2. 8400 citizens attended a concert. Let x, y and z be the number of children, the number of adults and the number of elderly people attending the concert respectively. It is also given that the sum of the number of children and the number of elderly people is less than the number of adults by 6288.

   (a) (i) Write down the first equation in x, y and z.

       (ii) Show that $x - y + z = -6288$.

   [2]

   The cost of a child ticket, an adult ticket and an elderly ticket were set to be $42, $84 and $21 respectively. The total cost of tickets sold in the concert was $655872.

   (b) Write down the third equation in x, y and z.

   [1]

   (c) Hence, find the values of x, y and z.

   [4]

Your Practice Set – Applications and Interpretation for IBDP Mathematics

3. The total price P of selling x apples, y oranges and z pears can be modelled by $P = ax + by + cz$, where a, b and c are the unit prices of apples, oranges and pears respectively. The following table shows the total prices of different combinations of apples, oranges and pears:

$P	x	y	z
$150	10	12	13
$178	14	8	19
$230	22	23	7

   (a) (i) Write down the system of three equations in a, b and c.

   (ii) Hence, find the values of a, b and c.

   [5]

   (b) Find the total price of 30 apples and 35 pears.

   [2]

4. In a football league, each team gains x points for a win, y points for a draw and z points for a loss. There are 46 matches in a year. The following table shows the performances of three teams in 2019:

Team	Win	Draw	Loss	Points
A	30	16	0	152
B	23	15	8	114
C	11	17	18	60

   (a) (i) Write down the system of three equations in x, y and z.

   (ii) Hence, find the values of x, y and z.

   [5]

   (b) Interpret the meaning of z.

   [1]

Chapter 7

Arithmetic Sequences

SUMMARY POINTs

✓ Properties of an arithmetic sequence u_n:
1. u_1: First term
2. $d = u_2 - u_1 = u_n - u_{n-1}$: Common difference
3. $u_n = u_1 + (n-1)d$: General term (n th term)
4. $S_n = \dfrac{n}{2}[2u_1 + (n-1)d] = \dfrac{n}{2}[u_1 + u_n]$: The sum of the first n terms

✓ $\sum_{r=1}^{n} u_r = u_1 + u_2 + u_3 + \cdots + u_{n-1} + u_n$: Summation sign

Solutions of Chapter 7

Your Practice Set – Applications and Interpretation for IBDP Mathematics

Paper 1 – Finding u_n and S_n with Given Terms

Example

In an arithmetic sequence, $u_1 = 3$ and $u_4 = 15$.

(a) Find d.

[2]

(b) Find u_{30}.

[2]

(c) Find S_{30}.

[2]

Solution

(a) $d = \dfrac{u_4 - u_1}{3}$ (M1) for finding d

$d = \dfrac{15 - 3}{3}$

$d = 4$ A1   N2

[2]

(b) $u_{30} = u_1 + (30-1)d$

$u_{30} = 3 + (30-1)(4)$ (A1) for correct substitution

$u_{30} = 119$ A1   N2

[2]

(c) $S_{30} = \dfrac{30}{2}\left[2u_1 + (30-1)d\right]$

$S_{30} = \dfrac{30}{2}\left[2(3) + (30-1)(4)\right]$ (A1) for correct substitution

$S_{30} = 1830$ A1   N2

[2]

Exercise 17

1. In an arithmetic sequence, $u_1 = 27$ and $u_5 = -1$.

    (a) Find d.
    [2]

    (b) Find u_{25}.
    [2]

    (c) Find S_{25}.
    [2]

2. In an arithmetic sequence, $u_1 = 3.5$ and $u_7 = 6.5$.

    (a) Find d.
    [2]

    (b) Find u_{42}.
    [2]

    (c) Find S_{84}.
    [2]

3. In an arithmetic sequence, the second term is 0 and the tenth term is 24.

    (a) Find the common difference.
    [2]

    (b) Find the fourth term.
    [2]

    (c) Find the sum of the first ten terms of the sequence.
    [2]

4. In an arithmetic sequence, the third term is $-\dfrac{2}{3}$ and the eighth term is $-\dfrac{22}{3}$.

    (a) Find the common difference.
    [2]

    (b) Find the eleventh term.
    [2]

    (c) Find the sum of the first forty terms of the sequence.
    [3]

Your Practice Set – Applications and Interpretation for IBDP Mathematics

18 Paper 1 – Real Life Problems

Example

Josie pays money into a saving scheme each year, for 50 years. In the first year she pays $4100, and her payments increases by the same amount each year. It is given that she pays $4350 in the second year.

(a) Find the amount that Josie will pay in the 15th year.

[3]

(b) Find the exact total amount that Josie will pay in over the 50 years.

[3]

Solution

(a) $d = 4350 - 4100$ (M1) for finding d
$d = 250$
The required amount
$= u_{15}$ (M1) for valid approach
$= 4100 + (15-1)(250)$
$= \$7600$ A1   N3

[3]

(b) The total amount
$= S_{50}$ (M1) for valid approach
$= \dfrac{50}{2}\bigl[2(4100) + (50-1)(250)\bigr]$ (A1) for substitution
$= \$511250$ A1   N3

[3]

Exercise 18

1. The lengths of the sides of a 20-sided polygon form an arithmetic sequence. The two shortest sides are 1.5 m and 1.9 m respectively.

   (a) Find the length of the longest side of the polygon.
   [3]

   (b) Find the perimeter of the polygon.
   [3]

2. Consider the following sequence of figures.

   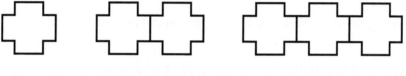

   Figure 1   Figure 2   Figure 3

   Figure 1 contains 12 edges and Figure 2 contains 23 edges.

   (a) Given that Figure n contains 221 edges, find n.
   [3]

   (b) Find the total number of edges of Figure 18, Figure 19 and Figure 20.
   [3]

3. Workers are drilling for oil in a large-scale project in Saudi Arabia. The costs of oil drilling to a depth of 10 m and a depth of 20 m are $7500 and $9300 respectively. It costs an extra constant amount for every subsequent extra depth of 10 m.

   (a) Find the cost of drilling to a depth of 100 m.
   [3]

   (b) The budget of the large-scale project is $340000. Find the greatest possible depth to be drilled.
   [4]

4. In the theatre of a town hall, there are 24 rows in total. The number of seats in each row forms an arithmetic sequence. The number of seats in the second row and the third row are 30 and 32 respectively.

   (a) Find the number of seats in the 24th row.
   [3]

   (b) The price of a ticket in the theatre is $75. Find the total income of the theatre if all tickets are sold.
   [4]

Your Practice Set – Applications and Interpretation for IBDP Mathematics

Paper 2 – Miscellaneous Problems

Example

In an arithmetic sequence, $S_{30} = 5430$ and $u_3 = 31$. Let u_1 and d be the first term and the common difference of the arithmetic sequence respectively.

(a) (i) Write down the first equation connecting u_1 and d.

(ii) Write down the second equation connecting u_1 and d.

(iii) Hence, write down the values of u_1 and d.

[6]

(b) Find u_n in terms of n.

[2]

(c) Find the smallest value of n such that $u_n > 888$.

[3]

(d) Find S_n in terms of n.

[2]

Solution

(a) (i) $S_{30} = 5430$

$\frac{30}{2}[2u_1 + (30-1)d] = 5430$ (M1) for valid approach

$30u_1 + 435d = 5430$ A1   N2

(ii) $u_3 = 31$

$u_1 + (3-1)d = 31$ (M1) for valid approach

$u_1 + 2d = 31$ A1   N2

(iii) $u_1 = 7$ A1   N1

$d = 12$ A1   N1

[6]

(b) $u_n = 7 + (n-1)(12)$ (M1) for valid approach

$u_n = 12n - 5$ A1   N2

[2]

58

SE Production Limited

(c) $u_n > 888$

$12n - 5 > 888$ (M1) for setting inequality

$12n > 893$

$n > \dfrac{893}{12}$ (A1) for correct value

Thus, the smallest value of n is 75.   A1   N3

[3]

(d) $S_n = \dfrac{n}{2}\left[2(7) + (n-1)(12)\right]$ (M1) for valid approach

$S_n = \dfrac{n}{2}(12n + 2)$

$S_n = 6n^2 + n$ A1   N2

[2]

Exercise 19

1. In an arithmetic sequence, $u_4 = 43$ and $S_{80} = -5320$. Let u_1 and d be the first term and the common difference of the arithmetic sequence respectively.

(a)   (i)   Write down the first equation connecting u_1 and d.

        (ii)   Write down the second equation connecting u_1 and d.

        (iii)   Hence, write down the values of u_1 and d.

[6]

(b)   Find u_n in terms of n.

[2]

(c)   Find the greatest value of n such that $u_n > 0$.

[3]

(d)   Find S_n in terms of n.

[2]

(e)   Find the value of n such that $S_n = -4425$.

[3]

Your Practice Set – Applications and Interpretation for IBDP Mathematics

2. In an arithmetic sequence, $u_{11} = 17$ and $S_{96} = 4512$. Let u_1 and d be the first term and the common difference of the arithmetic sequence respectively.

   (a) (i) Write down the first equation connecting u_1 and d.

   (ii) Write down the second equation connecting u_1 and d.

   (iii) Hence, write down the values of u_1 and d.
   [6]

   (b) Find u_n in terms of n.
   [2]

   (c) Find the greatest value of n such that $u_n < 147$.
   [3]

   (d) Find $u_{49} + u_{50} + u_{51} + \cdots + u_{95} + u_{96}$.
   [3]

   (e) Find the value of n such that $S_n = 4949$.
   [3]

3. In an arithmetic sequence, the general term is given by $u_n = 63 - 3n$, and the sum of the first n terms is given by S_n.

   (a) (i) Write down the values of u_1 and u_2.

   (ii) Hence, write down the common difference of the sequence.
   [3]

   (b) Find the smallest value of n such that $u_n \leq -21$.
   [3]

   (c) (i) Write down the values of S_1.

   (ii) Find S_n in terms of n.
   [3]

   (d) Find the value of n such that $S_n = 0$.
   [3]

   (e) Find the value of $\sum_{r=11}^{20} u_r$.
   [3]

4. In an arithmetic sequence, the general term is given by u_n, and the sum of the first n terms is given by $S_n = \dfrac{n^2 + 19n}{28}$.

    (a) (i) Write down the values of S_1 and S_2.

    (ii) Hence, write down the values of u_1 and u_2.

    (iii) Find the common difference of the sequence.

    [6]

    (b) Find the smallest value of n such that $\sum_{r=1}^{n} u_r > 100$.

    [3]

    (c) Find the value of $\sum_{r=50}^{100} u_r$.

    [3]

    (d) Find u_n in terms of n.

    [2]

    (e) Find the value of n such that $u_n + 2u_{n+1} = \dfrac{58}{7}$.

    [3]

Chapter 

Geometric Sequences

SUMMARY POINTs

✓ Properties of a geometric sequence u_n:
1. u_1: First term
2. $r = u_2 \div u_1 = u_n \div u_{n-1}$: Common ratio
3. $u_n = u_1 \times r^{n-1}$: General term (n th term)
4. $S_n = \dfrac{u_1(1-r^n)}{1-r}$: The sum of the first n terms

 Solutions of Chapter 8

20 | Paper 1 – Finding u_n and S_n with given terms

Example

The first three terms of a geometric sequence are 54, 18 and 6.

(a) Write down the value of r, the common ratio of the geometric sequence.

[1]

(b) Find u_7.

[2]

(c) Find the sum of the first 10 terms of this sequence, giving the answer correct to four decimal places.

[3]

Solution

(a) $r = \dfrac{1}{3}$ A1 N1

[1]

(b) $u_7 = u_1 \cdot r^{7-1}$

$u_7 = 54 \cdot \left(\dfrac{1}{3}\right)^{7-1}$ (A1) for substitution

$u_7 = \dfrac{2}{27}$ A1 N2

[2]

(c) $S_{10} = \dfrac{u_1(1-r^{10})}{1-r}$ (M1) for valid approach

$S_{10} = \dfrac{54\left(1-\left(\dfrac{1}{3}\right)^{10}\right)}{1-\dfrac{1}{3}}$ (A1) for substitution

$S_{10} = 80.99862826$

$S_{10} = 80.9986$ A1 N3

[3]

Your Practice Set – Applications and Interpretation for IBDP Mathematics

Exercise 20

1. The first three terms of a geometric sequence are 1024, 256 and 64.

    (a) Write down the value of r, the common ratio of the geometric sequence.
    [1]

    (b) Find u_8.
    [2]

    (c) Find the sum of the first twelve terms of this sequence, giving the answer correct to the nearest integer.
    [3]

2. The first three terms of a geometric sequence are $u_1 = 576$, $u_2 = 768$ and $u_3 = 1024$.

    (a) Write down the value of r.
    [1]

    (b) Find the value of $\sum_{n=1}^{7} u_n$, giving the answer correct to the nearest integer.
    [3]

    (c) Find the value of n such that $243u_n = 1048576$.
    [3]

3. The first three terms of a geometric sequence are $u_1 = 1.024$, $u_2 = 1.28$ and $u_3 = 1.6$.

    (a) Find the value of r.
    [2]

    (b) Find the least value of n such that $u_n > 5$.
    [3]

    (c) Find the value of $\sum_{n=1}^{10} u_n$.
    [2]

4. The first three terms of a geometric sequence are $u_1 = 1.5$, $u_2 = 2.4$ and $u_3 = 3.84$.

    (a) Find the value of r.
    [2]

    (b) Find the value of the eighth term of the sequence, giving the answer correct to two decimal places.
    [2]

    (c) Find the greatest value of n such that $S_n < 100$.
    [3]

21 Paper 1 – Real Life Problems

Example

A population of bees is growing at a rate of 6% per year. There were 160 bees at the beginning of 2002.

(a) Find the number of bees at the beginning of 2006.
[2]

(b) Find the number of bees n complete years after the beginning of 2002.
[2]

(c) It is assumed that each bee can produce 100 units of honey per year. Find the total amount of honey produced in a complete six-year period after the beginning of 2002.
[3]

Solution

(a) The number of bees
$= u_5$ (M1) for valid approach
$= 160 \times (1+6\%)^{5-1}$
$= 201.9963136$
$= 202$ A1   N2
[2]

(b) The number of bees
$= u_n$ (M1) for valid approach
$= 160 \times (1+6\%)^{n-1}$
$= 160 \times 1.06^{n-1}$ A1   N2
[2]

(c) The total amount of honey
$= 100 S_6$ (M1) for valid approach
$= 100 \left(\dfrac{160(1-1.06^6)}{1-1.06} \right)$ (A1) for substitution
$= 111605.0966$
$= 112000$ units   A1   N3
[3]

Your Practice Set – Applications and Interpretation for IBDP Mathematics

Exercise 21

1. The lengths of the sides of a 10-sided polygon form a geometric sequence. The three shortest sides are 1 m, 1.1 m and 1.21 m respectively.

    (a) Write down the exact length of the fourth shortest side of the polygon.
    [1]

    (b) Find the length of the longest side of the polygon.
    [2]

    (c) Find the perimeter of the polygon.
    [3]

2. In the theatre of a town hall, there are 12 rows in total. Each row consists of 30 seats. The price of a seat in the first row is $100. The prices of a seat in the following rows continue to decrease in the same ratio, such that the price of a seat in the second row is $90, and the price of a seat in the third row is $81, and so on. Give the answers correct to the nearest dollar in this question.

    (a) Find the price of a seat in the sixth row.
    [2]

    (b) Find the price of a seat in the nth row.
    [2]

    (c) Find the total income of the theatre if all tickets are sold.
    [3]

3. In a souvenir market in Khabarovsk located in East Russia, a set of 20 wooden dolls is displayed in a row and arranged in descending doll sizes. The volume of the largest doll is 24000 cm^3. The volume of the second largest doll is 95% of the previous doll, and this pattern continues.

    (a) Find the total volume of all wooden dolls when they are displayed in a single row.
    [3]

    (b) Find the number of wooden dolls with volume less than 10000 cm^3.
    [3]

4. A car is travelling in a journey. On 1st February the car travels 120 km. The distance travelled is expected to decrease by 10% each day.

    (a) Find the distance travelled on the fifth day.
    [2]

    (b) Find the difference between the distance travelled on the fifth day and the one on the sixth day.
    [2]

    (c) The car needs to refuel for every 500 km travelled. Find the date when the car refuels for the second time of this journey.
    [3]

22. Paper 2 – Arithmetic and Geometric Sequences

Example

Consider an arithmetic sequence $u_1, u_2, u_3, \ldots, u_n, \ldots$ where $u_1 = 264$, $u_2 = 276$ and $u_3 = 288$.

(a) Find the value of u_{17}.

[2]

(b) Find the sum of the first 23 terms of the sequence.

[3]

Now consider the sequence $v_1, v_2, v_3, \ldots, v_n, \ldots$ where $v_1 = \dfrac{40}{729}$, $v_2 = \dfrac{40}{243}$ and $v_3 = \dfrac{40}{81}$. This sequence continues in the same manner.

(c) Find the value of v_{13}.

[2]

(d) Find the sum of the first 13 terms of this sequence, giving the answer correct to the nearest integer.

[3]

Let k be a positive integer such that $u_k = v_k$.

(e) Calculate the value of k.

[3]

Solution

(a) u_{17}
$= u_1 + (17-1)d$ (M1) for valid approach
$= 264 + (17-1)(12)$
$= 456$ A1   N2

[2]

(b) The sum of the first 23 terms
$= \dfrac{23}{2}\left[2u_1 + (23-1)d\right]$ (M1) for valid approach

$= \dfrac{23}{2}\left[2(264) + (23-1)(12)\right]$ (A1) for substitution

$= 9108$ A1   N2

[3]

(c) v_{13}

$= v_1 \times r^{13-1}$ (M1) for valid approach

$= \dfrac{40}{729} \times 3^{13-1}$

$= 29160$ A1   N2

[2]

(d) The sum of the first 13 terms

$= \dfrac{v_1(1-r^{13})}{1-r}$ (M1) for valid approach

$= \dfrac{\dfrac{40}{729}(1-3^{13})}{1-3}$ (A1) for substitution

$= 43739.97257$

$= 43740$ A1   N2

[3]

(e) $u_k = v_k$

$264 + (k-1)(12) = \dfrac{40}{729} \times 3^{k-1}$ (M1) for setting equation

$264 + 12k - 12 = \dfrac{40}{729} \times 3^{k-1}$

$252 + 12k - \dfrac{40}{729} \times 3^{k-1} = 0$ (A1) for correct equation

By considering the graph of

$y = 252 + 12k - \dfrac{40}{729} \times 3^{k-1}$, $k = 9$.   A1   N2

[3]

Exercise 22

1. Kensuke purchased a new car for 24000 EUR on 1st January, 2011 and insures his car with an insurance company. In 2011, Kensuke needs to pay 1200 EUR for the insurance premium, and the amount he needs to pay is reduced by 15 EUR per year.

    (a) Find the amount of insurance premium Kensuke has paid in 2014.

    [2]

    The insurance company also estimates the value of Kensuke's car in each year. The company estimates that the car depreciates by 15% each year, such that the value of his car is 20400 EUR in 2012.

    (b) Find the exact value of the car in 2016.

    [2]

    (c) Find the year when the value of the car is first below 8000 EUR.

    [3]

    Kensuke will stop insuring his car if the amount of insurance premium is greater than the value of the car in a particular year.

    (d) Find the year when Kensuke stops insuring his car.

    [3]

    (e) Hence, find the total amount of insurance premium Kensuke has paid in this period.

    [3]

2. The table below shows the first four terms of four sequences: t_n, u_n, v_n and w_n.

n	1	2	3	4
t_n	50	100	200	400
u_n	50	70	110	170
v_n	50	1050	2050	3050
w_n	50	50	50	50

    (a) State the sequence which is

    (i) arithmetic;

    (ii) geometric;

    (iii) neither arithmetic nor geometric;

    (iii) both arithmetic and geometric.

    [4]

(b) For the arithmetic sequence,

    (i) Find the value of the 100th term.

    (ii) Find the sum of the first 25 terms.

[5]

(c) For the geometric sequence,

    (i) Find the value of the 7th term.

    (ii) Find the sum of the first 14 terms.

[5]

(d) The first m terms of the arithmetic sequence are not less than the first m terms of the geometric sequence, find the greatest value of m.

[3]

3. Giselle joins a training program in a sports centre. In the first training session Giselle runs a distance of 2400 metres. She increases her running distance by 200 metres after each training session.

(a) Find Giselle's running distance in the n th training session.

[2]

(b) In the x th training session, Giselle will first run further than 10 kilometres. Find the value of x.

[3]

(c) Find the total running distance in the first sixteen training sessions, giving the answer in the form $a \times 10^k$ km.

[3]

Giselle's best friend, Helena, followed Giselle to join the same training program, but due to different fitness conditions their running distances in each session can be different. In the first training session Helena runs a distance of 2000 metres. She increases her running distance by 5% after each training session.

(d) Find Helena's running distance in the 10th training session.

[2]

(e) After the w th training session, the total running distance of Giselle in the training program is less than that of Helena at the first time. Find the value of w.

[4]

4. A new coffee shop starts its business on 1st March, 2017. Its profit is EUR 2000 in the first month and increases by 15% every month.

   (a) Find the coffee shop's profit in November 2017.

   [2]

   (b) Find the coffee shop's total profit in 2017.

   [3]

   On 1st March, 2017 a new fast food shop nearby starts its business as well. Its profit is EUR 4000 in the first month and increases by EUR 1100 every month.

   (c) Find the fast food shop's profit in June 2018.

   [2]

   (d) The fast food shop's profit first exceeds EUR 30000 in the m th month. Find the value of m.

   [3]

   (e) The fast food shop's total profit is less than the coffee shop's total profit for the first time in the n th month. Find the value of n.

   [4]

Your Practice Set – Applications and Interpretation for IBDP Mathematics

Chapter

Financial Mathematics

SUMMARY POINTs

- ✓ Compound Interest:
  PV : Present value
  $r\%$: Interest rate per annum (per year)
  n: Number of years
  k: Number of compounded periods in one year
  $FV = PV\left(1+\dfrac{r}{100k}\right)^{kn}$: Future value
  $I = FV - PV$: Interest

- ✓ Inflation:
  $i\%$: Inflation rate
  $R\%$: Interest rate compounded yearly
  $(R-i)\%$: Real rate

SUMMARY POINTs

✓ Annuity:
  1. Payments at the beginning of each year

  2. Payments at the end of each year

     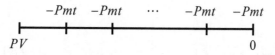

✓ Amortization:
  1. Payments at the beginning of each year

     $-Pmt \quad -Pmt \quad -Pmt \quad \cdots \quad -Pmt$

     $PV \qquad \qquad \qquad \qquad \qquad 0$

  2. Payments at the end of each year

     $-Pmt \quad -Pmt \quad \cdots \quad -Pmt \quad -Pmt$

     $PV \qquad \qquad \qquad \qquad \qquad 0$

 Solutions of Chapter 9

Your Practice Set – Applications and Interpretation for IBDP Mathematics

 Paper 1 – Compound Interest

Example

For this question, give all the answers correct to the nearest USD.

24000 USD is invested for 5 years at a nominal annual interest rate of 6%, compounded yearly.

(a) Find the value of P, the amount of money after 5 years.

[3]

(b) An amount of money Q is invested for 5 years at a nominal annual interest rate of 6%, compounded half-yearly. The amount of money after 5 years is P. Find the value of Q.

[3]

Solution

(a) $P = 24000\left(1 + \dfrac{6}{100}\right)^5$ (M1)(A1) for substitution

$P = 32117.41386$

$P = 32117$ USD A1 N3

By TVM Solver:
$N = 5$
$I\% = 6$
$PV = -24000$
$PMT = 0$ (M1)(A1) for correct values
$FV = ?$
$P/Y = 1$
$C/Y = 1$
$PMT : END$

$P = 32117$ USD A1 N3

[3]

(b) $Q\left(1 + \dfrac{6}{(100)(2)}\right)^{(2)(5)} = 32117.41386$ (M1)(A1) for correct equation

$Q(1.03)^{10} = 32117.41386$

$Q = 23898.37222$

$Q = 23898$ USD A1 N3

By TVM Solver:
N = 5
I% = 6
PV = ?
PMT = 0
FV = 32117.41386
P/Y = 1
C/Y = 2
PMT : END

$Q = 23898$ USD   (M1)(A1) for correct values

A1   N3

[3]

Exercise 23

1. For this question, give all the answers correct to the nearest 100 EUR.

   Aaron invested 360000 EUR in an account that pays a nominal annual interest rate of 3%, compounded half-yearly.

   (a) Find the value of P, the amount of money after 8 years.

   [3]

   (b) An amount of money Q is invested for 8 years at a nominal annual interest rate of 3%, compounded monthly. The amount of money after 8 years is P. Find the value of Q.

   [3]

2. $125000 is invested for 12 years at a nominal annual interest rate of 8%, compounded yearly.

   (a) Find the amount of money after 12 years, give the answer correct to the nearest 1000 dollars.

   [3]

   (b) Find the minimum number of complete years required for the amount of money to be doubled.

   [3]

3. P is invested for 5 years at a nominal annual interest rate of 4%, compounded quarterly. The amount of money after 5 years is 87000 AUD.

   (a) Find the value of P, give the answer correct to the nearest AUD.

   [3]

   (b) Find the minimum number of complete years required for the amount of money to be $2.5P$.

   [3]

4. $640000 is invested for t_1 years at a nominal annual interest rate of 5%, compounded half-yearly. The amount after t_1 years is doubled.

   Also, $640000 is invested for t_2 years at a nominal annual interest rate of 5%, compounded quarterly. The amount after t_2 years is doubled.

   Find the value of $t_1 - t_2$.

   [6]

24 Paper 1 – Equivalent Rates

Example

360000 USD is invested for 6 years at a nominal annual interest rate of 4%, compounded yearly.

(a) Find the value of P, the amount of money after 6 years, give the answer correct to the nearest USD.

[3]

(b) 360000 USD is invested for 6 years at a nominal annual interest rate of $r\%$, compounded half-yearly. The amount of money after 6 years is P. Find the value of r.

[3]

Solution

(a) $P = 360000\left(1 + \dfrac{4}{100}\right)^6$ (M1)(A1) for substitution

$P = 455514.8467$

$P = 455515$ USD   A1   N3

By TVM Solver:
N = 6
I% = 4
PV = −360000
PMT = 0
FV = ?
P/Y = 1
C/Y = 1
PMT : END

(M1)(A1) for correct values

$P = 455515$ USD   A1   N3

[3]

(b) $360000\left(1 + \dfrac{r}{(100)(2)}\right)^{(2)(6)} = 455514.8467$ (M1)(A1) for correct equation

$360000\left(1 + \dfrac{r}{200}\right)^{12} - 455514.8467 = 0$

By considering the graph of

$y = 360000\left(1 + \dfrac{r}{200}\right)^{12} - 455514.8467$,

$r = 3.9607805$.

Thus, $r = 3.96$. A1 N3

```
By TVM Solver:
N = 6
I% = ?
PV = -360000
PMT = 0
FV = 455514.8467
P/Y = 1
C/Y = 2
PMT : END
```
(M1)(A1) for correct values

Thus, $r = 3.96$. A1 N3

[3]

Exercise 24

1. Ben invested 54000 EUR in an account that pays a nominal annual interest rate of 6%, compounded monthly.

   (a) Find the value of P, the amount of money after 10 years, give the answer correct to the nearest 1000 EUR.

   [3]

   (b) 54000 EUR is invested for 10 years at a nominal annual interest rate of $r\%$, compounded quarterly. The amount of money after 10 years is P. Find the value of r.

   [3]

2. An amount of money is invested for 7 years at a nominal annual interest rate of 9%, compounded half-yearly. The amount of money after 7 years is $1600000.

   (a) Find the value of P, the original amount of money invested, give the answer correct to the nearest 10000 dollars.

   [3]

   (b) P is invested for n years at a nominal annual interest rate of 9%, compounded yearly. The amount of money after 7 years is $1600000. Find the value of n.

   [3]

3. Two equal amounts of money are invested in two different bank accounts, one at a nominal annual interest rate of 12%, compounded quarterly, for 4 years, with the another one at a nominal annual interest rate of 12%, compounded monthly, for n years. The amounts in both accounts at the end of the investments are equal. Find the value of n.

   [4]

4. Two equal amounts of money are invested in two different bank accounts, one at a nominal annual interest rate of 5%, compounded half-yearly, for 8 years, with the another one at a nominal annual interest rate of 5%, compounded k times per year, for 7.98 years. The amounts in both accounts at the end of the investments are equal. Find the value of k.

   [4]

25 Paper 1 – Rate of Inflation

Example

$170000 is invested for 8 years at a nominal annual interest rate of $r\%$, compounded yearly. The amount of money after 8 years is $260000.

(a) Find the value of r.

[3]

It is given that the rate of inflation during these 8 years is 2% per year.

(b) Write down the value of the real interest rate.

[1]

(c) Hence, find the real value of amount of money after 8 years.

[2]

Solution

(a) $260000 = 170000\left(1+\dfrac{r}{100}\right)^8$ (M1)(A1) for correct equation

$170000\left(1+\dfrac{r}{100}\right)^8 - 260000 = 0$

By considering the graph of

$y = 170000\left(1+\dfrac{r}{100}\right)^8 - 260000$,

$r = 5.454606$.

Thus, $r = 5.45$.   A1   N3

By TVM Solver:
N = 8
I% = ?
PV = −170000
PMT = 0   (M1)(A1) for correct values
FV = 260000
P/Y = 1
C/Y = 1
PMT : END

Thus, $r = 5.45$.   A1   N3

[3]

(b) 3.45%   A1   N1

[1]

(c) The real value of amount of money

Your Practice Set – Applications and Interpretation for IBDP Mathematics

$$= 170000\left(1 + \frac{3.454606}{100}\right)^8$$ (A1) for substitution

$= 223073.2887$

$= \$223000$ A1   N2

```
By TVM Solver :
N = 8
I% = 3.454606
PV = -170000
PMT = 0
FV = ?
P/Y = 1
C/Y = 1
PMT : END
```
(A1) for correct values

Thus, the real value is $223000.   A1   N2

[2]

Exercise 25

1. An amount of money $\$P$ is invested for 4 years at a nominal annual interest rate of 7%, compounded yearly. The amount of money after 4 years is $300000.

   (a) Find the value of P, the original amount of money invested.

   [3]

   It is given that the rate of inflation during these 4 years is 1.6% per year.

   (b) Write down the value of the real interest rate.

   [1]

   (c) Hence, find the real value of amount of money after 4 years.

   [2]

2. Ciana invested 8500 EUR in an account that pays a nominal annual interest rate of 12%, compounded monthly. This amount is invested for 9 years and the inflation rate in these 9 years is 1.8%.

   (a) Find the real interest rate per year.

   [4]

   (b) Find the real value of amount of interest incurred after 9 years.

   [3]

3. An amount of money $2800 is invested for 12 years at a nominal annual interest rate of 4%, compounded yearly. After the rate of inflation is considered, the real value of amount of money after 12 years is $4000.

   (a) Find the value of the real interest rate per year.

   [3]

   (b) Find the rate of inflation per year.

   [2]

4. Debby invested 14500 USD in an account that pays a nominal annual interest rate of 9.2%, compounded quarterly. This amount is invested for 8 years and the inflation rate in these 8 years is $i\%$.

   (a) Find the real interest rate per year, giving the answer in terms of i and correct to 4 decimal places.

   [4]

   It is given that the real value of amount of money after 8 years is 18500 USD.

   (b) Using the answer in (a), find the value of i.

   [3]

Your Practice Set – Applications and Interpretation for IBDP Mathematics

26 Paper 1 – Annuities

Example

Kevin is going to create an annuity fund which will pay him a monthly allowance of 2000 USD for 40 years after he is retired. In the fund, interest is earned 6% per year, compounded yearly.

(a) Find the value of the annuity fund that has to be saved, giving the answer correct to the nearest USD.

[3]

(b) Find the amount of payment needed per year if Kevin wants to save his money in his fund for 30 years.

[3]

Solution

(a) By TVM Solver:

$N = 40 \times 12$
$I\% = 6$
$PV = ?$
$PMT = 2000$
$FV = 0$
$P/Y = 12$
$C/Y = 1$
$PMT : END$

(M1)(A1) for correct values

$PV = -370397.2097$

Thus, the value of the annuity fund that has to be saved is 370937 USD.   A1   N3

[3]

(b) By TVM Solver:

$N = 30$
$I\% = 6$
$PV = 0$
$PMT = ?$
$FV = 370937.2097$
$P/Y = 1$
$C/Y = 1$
$PMT : END$

(M1)(A1) for correct values

$PMT = -4691.951934$

Thus, the amount of payment needed per year is 4690 USD.   A1   N3

[3]

Exercise 26

1. An annuity fund is created and it requires regular payments at the beginning of every year for 20 years. The value of the fund at the end of 20 years is $60000 and the interest is earned 7.5% per year.

    (a) Find the value of the regular payment per year.
    [3]

    (b) If the value of the regular payment per year is increased by $500, find the number of years required for the investment.
    [3]

2. Simon wants to accumulate an amount of money in an education fund at the end of 10 years. He deposits $1000 at the end of each month in the first five years, and deposits $1500 at the end of each month in the last five years. The interest is earned 3% per year.

    (a) Find the value of the investment after five years.
    [3]

    (b) Find the value of the investment after ten years.
    [3]

3. Teddy wants to accumulate an amount of money in an investment plans. He deposits $300 at the end of March, June, September and December every year. The interest is earned 5% per year.

    (a) Let P be the value of the investment after fifteen years. Find the value of P.
    [3]

    (b) If Teddy wants to adjust the amount of deposit such that the value of the investment after thirty years is $3.5P$, find the new amount of deposit.
    [3]

4. Consider the following table of the payments of the annuities X and Y, at the beginning of each month:

	Annuity X	Annuity Y
1st to 8th year	100	p
9th to 16th year	200	p

    It is given that the values of the investment after sixteen years for both annuities are the same. The interest is earned 2.9% per year for both annuities.

    (a) Find the value of the investment after eight years for annuity X.
    [3]

    (b) Find the value of the investment after sixteen years for annuity X.
    [3]

    (c) Hence, find the value of p.
    [2]

27 Paper 2 – Amortization

Example

Young Jae is going to buy a flat. He is suggested to choose one of the two options to repay the loan of 250000 USD with a nominal annual interest rate of 9%:

Option 1: A total of 240 equal monthly payments have to be paid at the end of each month
Option 2: A monthly payment of 2000 USD has to be paid at the end of each month until the loan is fully repaid

(a) If Young Jae selects the option 1, find

  (i) the amount of monthly payment,

  (ii) the total amount to be paid,

  (iii) the amount of interest paid.
[7]

(b) If Young Jae selects the option 2, find

  (i) the number of months to repay the loan,

  (ii) the total amount to be paid,

  (iii) the amount of interest paid.
[7]

(c) By considering the amounts of monthly payment in both options, state the better option for Young Jae and explain the answer.
[2]

(d) By considering the amounts of interest paid in both options, state the better option for Young Jae and explain the answer.
[2]

Solution

(a) (i) By TVM Solver:

$$\begin{array}{|l|}\hline N = 240 \\ I\% = 9 \\ PV = 250000 \\ PMT = ? \\ FV = 0 \\ P/Y = 12 \\ C/Y = 1 \\ PMT : END \\ \hline\end{array}$$

(M1)(A1) for correct values

PMT = −2193.157954

Thus, the amount of monthly payment is 2190 USD. A1 N3

(ii) The total amount to be paid
= (2193.157954)(240) (M1) for valid approach
= 526357.909
= 526000 USD A1 N2

(iii) The amount of interest paid
= 526357.909 − 250000 (M1) for valid approach
= 276357.909
= 276000 USD A1 N2

[7]

(b) (i) By TVM Solver:

$$\begin{array}{|l|}\hline N = ? \\ I\% = 9 \\ PV = 250000 \\ PMT = -2000 \\ FV = 0 \\ P/Y = 12 \\ C/Y = 1 \\ PMT : END \\ \hline\end{array}$$

(M1)(A1) for correct values

N = 321.9090083

Thus, the number of months to repay the loan is 322 months. A1 N3

(ii) The total amount to be paid
= (2000)(322) (M1) for valid approach
= 644000 USD A1 N2

(iii) The amount of interest paid
= 644000 − 250000 (M1) for valid approach
= 394000 USD A1 N2

[7]

(c) The amount of monthly payment in option 2 is less than that in option 1. R1
Thus, the option 2 is better. A1 N0

[2]

(d) The amount of interest paid in option 1 is less than that in option 2. R1
Thus, the option 1 is better. A1 N0

[2]

Exercise 27

1. Takumi is going to purchase a boat. He is suggested to choose one of the two options to repay the loan of $1900000:

   Option 1: A total of 144 equal monthly payments have to be paid at the end of each month, with a nominal annual interest rate of 3.7%
   Option 2: A deposit of $350000 has to be paid at the beginning of the loan, followed by monthly payments of $17500 at the end of each month until the loan is fully repaid, with a nominal annual interest rate of 3.4%

   (a) If Takumi selects the option 1, find

   (i) the amount of monthly payment,

   (ii) the total amount to be paid,

   (iii) the amount of interest paid.

   [7]

   (b) If Takumi selects the option 2, find

   (i) the number of months to repay the loan,

   (ii) the exact total amount to be paid,

   (iii) the amount of interest paid.

   [7]

   (c) By considering the amounts of monthly payment in both options, state the better option for Takumi and explain the answer.

   [2]

(d) By considering the amounts of interest paid in both options, state the better option for Takumi and explain the answer.

[2]

2. Bosco needs to settle a payment for his postgraduate programme. He is suggested by a bank to choose one of the two options to repay the loan of $40000 with a nominal annual interest rate of 4.5%:

Option 1: A deposit of $10000 has to be paid at the beginning of the loan, followed by a total of 36 equal monthly payments at the end of each month
Option 2: A monthly payment of $800 has to be paid at the end of each month until the loan is fully repaid

(a) If Bosco selects the option 1, find

(i) the amount of monthly payment,

(ii) the amount of interest paid.

[6]

(b) If Bosco selects the option 2, find

(i) the number of months to repay the loan, rounding up the answer correct to the nearest month,

(ii) the amount of interest paid.

[6]

(c) By considering the amounts of monthly payment in both options, state the better option for Bosco and explain the answer.

[2]

(d) By considering the amounts of interest paid in both options, state the better option for Bosco and explain the answer.

[2]

The bank also suggested the option 3 to Bosco, such that a total of 60 equal monthly payments of $900 have to be paid at the end of each month, with a nominal annual interest rate of r%

(e) Find the value of r.

[3]

3. Clement works in a bank and he is designing the structure of an amortization schedule for customers. He suggests the version 1 of the amortization schedule for a loan of $10000, with a nominal annual interest rate of 2%:

Version 1: A total of 120 equal monthly payments of $R have to be paid at the end of each month in a ten-year period

(a) Find

    (i) the value of R,

    (ii) the amount of interest paid.

[6]

Clement then makes some amendments in the version 1, such that the version 2 of the amortization schedule is as follows:

Version 2: A total of 60 equal monthly payments of $R have to be paid at the end of each month in the first five years, then the amount of monthly payments of $(R+60) have to be paid at the end of each month, until the loan is fully repaid

(b) (i) Find the number of months to repay the loan, rounding up the answer correct to the nearest month.

    (ii) Find the amount of interest paid.

    (iii) Explain the reason why the version 2 of the amortization schedule is more favourable to customers than the version 1.

[9]

Clement later makes the final amendments in the version 2, such that the version 3 of the amortization schedule is as follows:

Version 3: A monthly payment of $1.5R has to be paid at the end of each month until the loan is fully repaid

(c) (i) Find the number of months to repay the loan, rounding up the answer correct to the nearest month.

    (ii) Hence, write down the difference of the number of months required to repay the loan between the version 2 and the version 3.

[4]

4. Daniel is a financial analyst and he is investigating the differences of various amortization schedules when some of the factors are changed.

Firstly, he is investigating the differences between the amortization schedules with payments at the beginning of each year and at the end of each year, by considering the following two versions:

Version 1: A total of 20 equal yearly payments of $$R_1$ have to be paid at the beginning of each year

Version 2: A total of 20 equal yearly payments of $$R_2$ have to be paid at the end of each year

Both versions are designed for a loan of $50000, with a nominal annual interest rate of 2%.

(a) (i) Find the value of R_1.

(ii) Find the value of R_2.

(iii) Interpret the meaning of $$20(R_2 - R_1)$.

(iv) State which version will have the smaller total amount to be paid.

[8]

In addition, he is also investigating the differences between the amortization schedules with yearly payments and monthly payments, by considering the following additional version:

Version 3: A total of 240 equal monthly payments of $$R_3$ have to be paid at the end of each month

The version 3 is designed for a loan of $50000, with a nominal annual interest rate of 2%.

(b) (i) Find the value of R_3.

(ii) Interpret the meaning of $$(240R_3 - 50000)$.

(iii) By comparing the version 2 and the version 3, determine which version will have the smaller total amount to be paid. Justify the answer.

[7]

Chapter 10

Coordinate Geometry

SUMMARY POINTs

- Consider the points $P(x_1, y_1)$ and $Q(x_2, y_2)$ on a x-y plane:
    1. $m = \dfrac{y_2 - y_1}{x_2 - x_1}$: Slope of PQ
    2. $d = \sqrt{(x_2 - x_1)^2 + (y_2 - y_1)^2}$: Distance between P and Q
    3. $\left(\dfrac{x_1 + x_2}{2}, \dfrac{y_1 + y_2}{2}\right)$: Mid-point of PQ

- Consider the points $P(x_1, y_1, z_1)$ and $Q(x_2, y_2, z_2)$ on a x-y-z plane:
    1. z-axis: The axis perpendicular to the x-y plane
    2. $d = \sqrt{(x_2 - x_1)^2 + (y_2 - y_1)^2 + (z_2 - z_1)^2}$: Distance between P and Q
    3. $\left(\dfrac{x_1 + x_2}{2}, \dfrac{y_1 + y_2}{2}, \dfrac{z_1 + z_2}{2}\right)$: Mid-point of PQ

> **SUMMARY POINTs**
>
> ✓ Forms of straight lines with slope m and y-intercept c:
>   1. $y = mx + c$: Slope-intercept form
>   2. $Ax + By + C = 0$: General form
>
> ✓ Ways to find the x-intercept and the y-intercept of a line:
>   1. Substitute $y = 0$ and make x the subject to find the x-intercept
>   2. Substitute $x = 0$ and make y the subject to find the y-intercept

 Solutions of Chapter 10

28 Paper 1 – Three-Dimensional Space

Example

In a three-dimensional space, the coordinates of the point P are $(4, 6, 12)$.

(a) Find the length of OP, where O is the origin.

[2]

(b) Find the coordinates of M, the mid-point of OP.

[2]

The coordinates of the point A are $(x, 15, 6)$. It is given that $MA = 13$.

(c) Find the possible values of x.

[3]

Solution

(a) The length of OP
$= \sqrt{(4-0)^2 + (6-0)^2 + (12-0)^2}$ (A1) for substitution
$= 14$ A1   N2

[2]

(b) The coordinates of M
$= \left(\dfrac{0+4}{2}, \dfrac{0+6}{2}, \dfrac{0+12}{2} \right)$ (A1) for substitution
$= (2, 3, 6)$ A1   N2

[2]

(c) $\sqrt{(x-2)^2 + (15-3)^2 + (6-6)^2} = 13$ (M1) for setting equation

$\sqrt{(x-2)^2 + 144} = 13$

$\sqrt{(x-2)^2 + 144} - 13 = 0$

By considering the graph of
$y = \sqrt{(x-2)^2 + 144} - 13$, $x = -3$ or $x = 7$.   A2   N3

[3]

Exercise 28

1. In a three-dimensional coordinate system with vertical z-axis, the coordinates of the point A are $(-14, -48, 0)$. O is the origin.

    (a) Find the coordinates of M, the mid-point of OA.

    [2]

    The point N is vertically above the point M by n units, where $n > 0$. It is given that ON = 65.

    (b) (i) Write down the coordinates of N, giving the answer in terms of n.

    (ii) Find the value of n.

    [3]

2. In a three-dimensional coordinate system with vertical z-axis, the coordinates of the point P and Q are $(15, 25, 35)$ and $(45, 85, 15)$ respectively.

    (a) Find the length of PQ.

    [2]

    (b) Find the coordinates of M, the mid-point of PQ.

    [2]

    The point N is vertically below the point M by 20 units.

    (c) (i) Write down the coordinates of N.

    (ii) Hence, find the length of QN.

    [3]

3. In a three-dimensional coordinate system with vertical z-axis, the coordinates of the point A and B are $(12, -2, -8)$ and $(-24, 30, 4)$ respectively.

    (a) Find the coordinates of M, the mid-point of AB.

    [2]

    M is translated 40 units in the positive x direction, following by another translation of 9 units in the positive y direction to reach the point N.

    (b) Find the length of MN.

    [2]

    N is translated upward by 30.75 units to reach the point P.

    (c) Find the angle of elevation of P from M.

    [2]

Your Practice Set – Applications and Interpretation for IBDP Mathematics

4. In a three-dimensional coordinate system with vertical z-axis, the coordinates of the point P and Q are $(-16, 8, 24)$ and $(29, -100, h)$ respectively, where $h < 24$. It is given that $PQ = 195$.

    (a) Find the value of h.

    [2]

    Q is translated upward by 156 units to reach the point R, such that the triangle PQR is a right-angled triangle.

    (b) Find the angle of depression of Q from P.

    [2]

    P is the mid-point of QS.

    (c) Find the coordinates of S.

    [2]

29 Paper 1 – Finding the Equation of a Straight Line

Example

A straight line L passes through the points $A(3, 2)$ and $B(10, 16)$.

(a) Find the equation of L, giving the answer in general form. [3]

(b) Write down the x-intercept and the y-intercept of L. [2]

(c) Write down the coordinates of the midpoint of AB. [1]

Solution

(a) The gradient of L
$$= \frac{16-2}{10-3}$$
$$= 2$$
(A1) for correct formula

The equation of L:
$y - 2 = 2(x - 3)$ (M1) for substitution
$y - 2 = 2x - 6$
$2x - y - 4 = 0$ A1   N3

[3]

(b) The x-intercept of L is 2   A1
The y-intercept of L is –4   A1   N2

[2]

(c) $(6.5, 9)$ A1   N1

[1]

Your Practice Set – Applications and Interpretation for IBDP Mathematics

Exercise 29

1. A straight line L passes through the points $(10, 6)$ and $(20, 11)$.

    (a) Find the equation of L, giving the answer in general form.
    [3]

    L cuts the x-axis and the y-axis at A and B respectively.

    (b) Write down the x-intercept and the y-intercept of L.
    [2]

    (c) Hence, write down the coordinates of the midpoint of AB.
    [1]

2. A straight line L passes through the points $(-4, -8)$ and $(2, -26)$.

    (a) Find the equation of L, giving the answer in general form.
    [3]

    (b) Write down the x-intercept and the y-intercept of L.
    [2]

    (c) Write down the slope of the straight line passing through $(-4, -8)$ and perpendicular to L.
    [1]

3. A straight line L_1 passes through the points $(5, 1)$ and $(17, 37)$.

    (a) Find the equation of L_1, giving the answer in general form.
    [3]

    (b) The equation of another straight line, L_2, is given as $3x - y - 100 = 0$. Find the geometric relationship between L_1 and L_2.
    [2]

4. A straight line L_1 passes through the points $(-4, 0)$ and $(4, 40)$.

    (a) Find the equation of L_1, giving the answer in general form.
    [3]

    (b) The equation of another straight line, L_2, is given as $x + 5y + 150 = 0$. Find the geometric relationship between L_1 and L_2.
    [2]

30 Paper 1 – Parallel and Perpendicular Lines

Example

The equation of a straight line L_1 is given as $2x + y - 10 = 0$.

(a) Write down the gradient and the x-intercept of L_1.

[2]

(b) Find the equation of another straight line L_2 such that L_2 is parallel to L_1 and L_2 passes through $(4, 8)$, giving the answer in general form.

[3]

Solution

(a) The gradient of L_1 is –2   A1

The x-intercept of L_1 is 5   A1   N2

[2]

(b) The gradient of L_2 is –2   (A1) for correct gradient

The equation of L_2:

$y - 8 = -2(x - 4)$ (M1) for substitution

$y - 8 = -2x + 8$

$2x + y - 16 = 0$ A1   N3

[3]

Exercise 30

1. The equation of a straight line L_1 is given as $x - 2y + 16 = 0$.

    (a) Write down the gradient and the y-intercept of L_1.

    [2]

    (b) Find the equation of another straight line L_2 such that L_2 is parallel to L_1 and L_2 passes through $(-2, 5)$, giving the answer in general form.

    [3]

Your Practice Set – Applications and Interpretation for IBDP Mathematics

2. The equation of a straight line L_1 is given as $3x + 2y - 4 = 0$.

   (a) Write down the gradient of L_1.

   [1]

   (b) L_1 passes through the point $(4, a)$. Find the value of a.

   [2]

   (c) Find the equation of another straight line L_2 such that L_2 is parallel to L_1 and L_2 passes through $(1, -7)$, giving the answer in general form.

   [3]

3. The equation of a straight line L_1 is given as $3x + y + 21 = 0$.

   (a) Write down the gradient and the x-intercept of L_1.

   [2]

   (b) Find the equation of another straight line L_2 such that L_2 is perpendicular to L_1 and they intersect on the x-axis, giving the answer in general form.

   [3]

4. The equation of a straight line L_1 is given as $2x - 4y - 17 = 0$.

   (a) (i) Write down the gradient of L_1.

   (ii) Write down the y-intercept of L_1.

   [2]

   (b) Find the equation of another straight line L_2 such that L_2 is perpendicular to L_1 and they intersect on the y-axis, giving the answer in general form.

   [3]

   (c) L_2 passes through the point $(b, 5.75)$. Find the value of b.

   [2]

31 Paper 2 – Miscellaneous Problems

Example

The straight line L_1 passes through the points $A(0, 3)$ and $B(6, 6)$.

(a) Find the gradient of L_1.

[2]

(b) Find the equation of L_1, giving the answer in general form.

[3]

Another straight line L_2 passes through the origin and is parallel to L_1.

(c) Write down the equation of L_2, giving the answer in slope-intercept form.

[2]

C is the mid-point of AB.

(d) Find the coordinates of C.

[2]

Another straight line L_3 passes through the point C and is perpendicular to L_1.

(e) Find the equation of L_3, giving the answer in general form.

[3]

(f) Find the x-intercept of L_3.

[2]

L_2 and L_3 intersects at the point D.

(g) Find the coordinates of D.

[3]

Solution

(a) The gradient of L_1

$$= \frac{6-3}{6-0}$$ (M1) for valid approach

$$= \frac{1}{2}$$ A1   N2

[2]

Your Practice Set – Applications and Interpretation for IBDP Mathematics

(b) The equation of L_1:

$y - 3 = \dfrac{1}{2}(x - 0)$ (M1) for substitution

$2y - 6 = x$ (A1) for simplification

$x - 2y + 6 = 0$ A1   N3

[3]

(c) The equation of L_2 is $y = \dfrac{1}{2}x$.   A2   N2

[2]

(d) The coordinates of C

$= \left(\dfrac{0+6}{2}, \dfrac{3+6}{2}\right)$ (A1) for substitution

$= (3, 4.5)$ A1   N2

[2]

(e) The gradient of L_3

$= -1 \div \dfrac{1}{2}$

$= -2$ (A1) for correct value

The equation of L_3:

$y - 4.5 = -2(x - 3)$ (M1) for substitution

$y - 4.5 = -2x + 6$

$2x + y - 10.5 = 0$

$4x + 2y - 21 = 0$ A1   N3

[3]

(f) $4x + 2(0) - 21 = 0$ (M1) for substitution

$x = 5.25$

Thus, the x-intercept of L_3 is 5.25. A1   N2

[2]

(g) $4x + 2\left(\dfrac{1}{2}x\right) - 21 = 0$ (M1) for substitution

$5x - 21 = 0$

$x = 4.2$

$y = \dfrac{1}{2}(4.2)$ (M1) for substitution

$y = 2.1$

Thus, the coordinates of D are $(4.2, 2.1)$. A1   N3

[3]

Exercise 31

1. The straight line L_1 passes through the points $A(-4, 6)$ and $B(-2, 0)$.

    (a) Find the gradient of L_1.

    [2]

    (b) Find the equation of L_1, giving the answer in general form.

    [3]

    C is a point on the x-axis such that $OB = OC$, where O is the origin. Another straight line L_2 passes through the point C and is parallel to L_1.

    (c) Find the equation of L_2, giving the answer in slope-intercept form.

    [3]

    D is the mid-point of AC.

    (d) Find the coordinates of D.

    [2]

    Another straight line L_3 passes through the point D and is perpendicular to L_1.

    (e) Find the equation of L_3, giving the answer in general form.

    [3]

    (f) It is given that $CD = \dfrac{k}{\sqrt{5}} BD$. Find the value of k.

    [4]

2. The straight line L_1 passes through the points $A(0, k)$ and $B(40, 30)$. $C(20, 25)$ is the mid-point of AB.

    (a) Find the value of k.

    [2]

    (b) Find the gradient of L_1.

    [2]

    (c) Find the equation of L_1, giving the answer in slope-intercept form.

    [2]

    Another straight line L_2 passes through the point A and is perpendicular to L_1.

    (d) Write down the equation of L_2, giving the answer in slope-intercept form.

    [3]

    Another straight line L_3 passes through the point $D(k, 0)$ and is perpendicular to L_1.

Your Practice Set – Applications and Interpretation for IBDP Mathematics

    (e)   Find the equation of L_3, giving the answer in general form.

[3]

    (f)   The point $E(r, r)$ lies on L_3. Find the value of r.

[2]

L_1 and L_3 intersects at the point F.

    (g)   Find the coordinates of F, giving the answers in exact values.

[3]

3.   The straight line L_1 passes through the points $A(3k, 0)$ and $B(0, 2k)$, where k is a positive constant.

    (a)   Find the gradient of L_1.

[2]

    (b)   Find the equation of L_1, giving the answer in general form and in terms of k.

[3]

The straight line L_1 also passes through the point $C(-30, 40)$.

    (c)   Find the value of k.

[2]

Another straight line L_2 passes through the point $D(1.5k, -2.5)$ and is perpendicular to L_1.

    (d)   Find the equation of L_2, giving the answer in general form.

[3]

Consider the graph of the quadratic function $f(x) = a(x-h)^2 + k$. The graph passes through the point B and its vertex is D.

    (e)   Write down the values of h and k.

[2]

    (f)   Find the value of a.

[2]

    (g)   Find the x-intercepts of the graph of $f(x)$.

[3]

4. The graph of the quadratic function $f(x) = 2x^2 + 4x - 16$ passes through $A(a, 0)$, $B(b, 0)$ and $C(0, c)$, where $a < b$. $V(h, k)$ is the vertex of the graph.

(a) Find the values of a and b.

[3]

(b) Write down the value of c.

[1]

(c) Write down the values of h and k.

[2]

(d) Find the gradient of VB.

[2]

(e) Find the equation of VB, giving the answer in general form.

[3]

The point $D(d, 0)$ lies on the x-axis. Find the value of d if

(f) (i) CD $\perp$ VB;

(ii) CD//VB.

[5]

Your Practice Set – Applications and Interpretation for IBDP Mathematics

Paper 2 – Areas Bounded by Straight Lines

Example

The straight line L_1 passes through the points $A(-15, 0)$ and $B(0, 20)$.

(a) Find the gradient of L_1.

[2]

(b) Find the equation of L_1, giving the answer in general form.

[3]

Another straight line L_2 passes through the origin O and is perpendicular to L_1.

(c) Find the equation of L_2, giving the answer in slope-intercept form.

[2]

L_1 and L_2 intersects at the point C.

(d) Find the coordinates of C.

[3]

(e) Find the area of the triangle OBC.

[2]

(f) Find the perimeter of the triangle OBC.

[4]

Solution

(a) The gradient of L_1

$= \dfrac{0-20}{-15-0}$ (M1) for valid approach

$= \dfrac{4}{3}$ A1   N2

[2]

(b) The equation of L_1:

$y - 20 = \dfrac{4}{3}(x - 0)$ (M1) for substitution

$3y - 60 = 4x$ (A1) for simplification

$4x - 3y + 60 = 0$ A1   N3

[3]

(c) The gradient of L_2

$= -1 \div \dfrac{4}{3}$

$= -\dfrac{3}{4}$ (A1) for correct value

The equation of L_2:

$y = -\dfrac{3}{4}x$ A1   N2

[2]

(d) $4x - 3\left(-\dfrac{3}{4}x\right) + 60 = 0$ (M1) for substitution

$6.25x = -60$

$x = -9.6$

$y = -\dfrac{3}{4}(-9.6)$ (M1) for substitution

$y = 7.2$

Thus, the coordinates of C are $(-9.6, 7.2)$.   A1   N3

[3]

(e) The area of the triangle OBC

$= \dfrac{(20-0)(0-(-9.6))}{2}$ (A1) for correct formula

$= 96$ A1   N2

[2]

(f) $BC = \sqrt{(-9.6-0)^2 + (7.2-20)^2}$

$BC = 16$ (A1) for correct value

$OC = \sqrt{(-9.6-0)^2 + (7.2-0)^2}$

$OC = 12$ (A1) for correct value

The perimeter of the triangle OBC

$= 12 + 16 + 20$ (M1) for valid approach

$= 48$ A1   N4

[4]

Your Practice Set – Applications and Interpretation for IBDP Mathematics

Exercise 32

1. The straight line L_1 passes through the points $A(0, 100)$ and $B(200, 0)$.

    (a) Find the gradient of L_1.

    [2]

    (b) Find the equation of L_1, giving the answer in general form.

    [3]

    Another straight line L_2 passes through the origin O and is perpendicular to L_1. The equation of L_2 is given by $y = mx + c$.

    (c) Write down the values of m and c.

    [2]

    L_1 and L_2 intersects at the point C.

    (d) Find the coordinates of C.

    [3]

    (e) Find the area of the triangle OAC.

    [2]

    (f) It is given that the ratio of the area of the triangle OBC to the area of the triangle OAC is $r : 1$. Find the value of r.

    [3]

2. In the diagram below, the straight line L_1 passes through the points $A(-60, 0)$ and $B(-20, 20)$. The straight line L_2 passes through the point $C(-30, 0)$ and is parallel to L_1. The straight line L_3 passes through the point B and is perpendicular to L_1. L_3 cuts the x-axis at the point E. L_2 and L_3 intersect at the point D.

    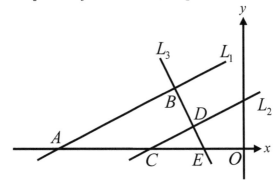

    (a) Find the gradient of L_1.

    [2]

(b) Find the equation of L_1, giving the answer in general form.

[3]

(c) Find the equation of L_2, giving the answer in slope-intercept form.

[2]

(d) Find the equation of L_3, giving the answer in slope-intercept form.

[3]

(e) Find the coordinates of D.

[3]

(f) Hence, find the area of the triangle CDE.

[4]

3. In the diagram below, the straight line L_1 passes through the points $A(0, a)$ and $B(30, 0)$. The gradient of L_1 is $-\dfrac{4}{3}$. Another straight line L_2 passes through the points B and $E(90, 45)$, and is perpendicular to L_1. The points $C(3c, c)$ and D lie on L_2 and AE respectively such that CD is parallel to L_1.

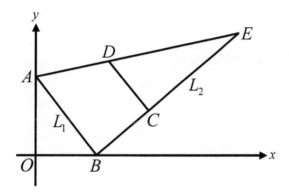

(a) Find the value of a.

[2]

(b) Find the equation of L_1, giving the answer in general form.

[3]

(c) Find the equation of L_2, giving the answer in slope-intercept form.

[3]

(d) Find the value of c.

[2]

It is also given that $DE = 15\sqrt{13}$ and $CD = 30$.

(e) Find the length of CE.

[3]

(f) Hence, find the area of the triangle CDE.

[2]

Your Practice Set – Applications and Interpretation for IBDP Mathematics

4. The straight line L_1 passes through the origin O and the points A(−20, 10) and B(p, q). The mid-point of AB is O.

   (a) Find the gradient of L_1.

   [2]

   (b) Write down the values of p and q.

   [2]

   (c) Find the equation of L_1, giving the answer in slope-intercept form.

   [2]

   Another straight line L_2 passes through the point C and the origin, and is perpendicular to L_1.

   (d) Find the equation of L_2, giving the answer in general form.

   [3]

   It is given that the area of the triangle ABC is 250, and C lies in the first quadrant.

   (e) Find the length of OC, giving the answer in surd form.

   [3]

   (f) Hence, find the coordinates of C.

   [4]

Chapter 11

Voronoi Diagrams

SUMMARY POINTs

✓ Elements in Voronoi Diagrams:
   Site: A given point
   Cell of a site: A collection of points which is closer to the site than other sites
   Boundary: A line dividing the cells
   Vertex: An intersection of boundaries

✓ Related problems:
   1. Nearest neighbor interpolation
   2. Incremental algorithm
   3. Toxic waste dump problem

Solutions of Chapter 11

Your Practice Set – Applications and Interpretation for IBDP Mathematics

Paper 1 – Equations of Perpendicular Bisectors

Example

Consider the points $A(4, 2)$ and $B(8, 8)$.

(a) Find the coordinates of C, the mid-point of AB.

[2]

(b) (i) Find the gradient of AB.

(ii) Hence, write down the gradient of the straight line perpendicular to AB.

[3]

(c) Find the equation of the perpendicular bisector of AB, giving the answer in slope-intercept form.

[2]

Solution

(a) The coordinates of C
$= \left(\dfrac{4+8}{2}, \dfrac{2+8}{2} \right)$ (A1) for substitution
$= (6, 5)$ A1   N2

[2]

(b) (i) The gradient of AB
$= \dfrac{8-2}{8-4}$ (M1) for valid approach
$= \dfrac{3}{2}$ A1   N2

(ii) $-\dfrac{2}{3}$ A1   N1

[3]

(c) The equation of the perpendicular bisector of AB:
$y - 5 = -\dfrac{2}{3}(x - 6)$ (M1) for substitution
$y - 5 = -\dfrac{2}{3}x + 4$
$y = -\dfrac{2}{3}x + 9$ A1   N2

[2]

Exercise 33

1. Consider the points $A(0, 6)$ and $B(0, 24)$.

    (a) Find the coordinates of C, the mid-point of AB.

[2]

    (b) Write down the gradient of the straight line perpendicular to AB.

[1]

    (c) Write down the equation of the perpendicular bisector of AB.

[1]

    (d) State the geometric relationship between A, B and D, where D is a point on the perpendicular bisector of AB.

[1]

2. Consider the straight line $x + 2y = 100$, with the x-intercept and the y-intercept at A and B respectively.

    (a)   (i) Write down the coordinates of A.

         (ii) Write down the coordinates of B.

         (iii) Hence, find the coordinates of C, the mid-point of AB.

[4]

The gradient of AB is -0.5.

    (b) Write down the gradient of the straight line perpendicular to AB.

[1]

    (c) Find the equation of the perpendicular bisector of AB, giving the answer in slope-intercept form.

[2]

Your Practice Set – Applications and Interpretation for IBDP Mathematics

3. The equation of the perpendicular bisector of two points A and B is $y = -\dfrac{4}{3}x + 35$, where the coordinates of B are $(12, -6)$.

   (a) (i) Write down the gradient of AB.

   (ii) Hence, find the equation of AB, giving the answer in slope-intercept form.

   [3]

   (b) Find the coordinates of the point of intersection of AB and the perpendicular bisector of AB.

   [3]

   (c) Find the coordinates of A.

   [2]

4. The equation of L, the perpendicular bisector of two points A and B, has the gradient $-\dfrac{3}{4}$, where the coordinates of A are $(0, 20)$.

   (a) (i) Write down the gradient of AB.

   (ii) Hence, find the equation of AB, giving the answer in slope-intercept form.

   [3]

   Let b be the x-coordinate of B, where b is a positive constant.

   (b) Write down the y-coordinate of B in terms of b.

   [1]

   The point C$(25, 20)$ lies on L.

   (c) (i) Write down the length of AC.

   (ii) Hence, find the value of b.

   [3]

34 Paper 1 – Nearest Neighbour Interpolation

Example

The diagram below shows the Voronoi diagram of three post offices, A, B and C, in a town, where 1 unit represents 1 km.

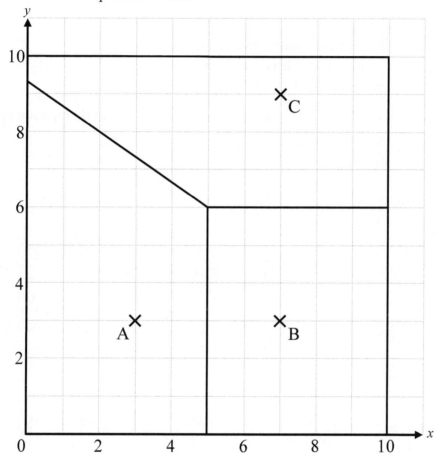

(a) Write down the area of the Voronoi cell of the post office B.

[1]

(b) (i) Find the mid-point of AC.

(ii) Hence, find the perpendicular distance between C and the perpendicular bisector of AC.

[4]

Clara would like to find a post office which is nearest to her home to send a parcel overseas. The position of her home is at $(1, 8)$.

(c) State which post office would be the best choice for her.

[1]

Your Practice Set – Applications and Interpretation for IBDP Mathematics

Solution

(a)   30 km²   A1   N1

[1]

(b) (i)   The mid-point of AC
$$= \left(\frac{3+7}{2}, \frac{3+9}{2}\right)$$ (A1) for substitution
$$= (5, 6)$$ A1   N2

(ii)   The required perpendicular distance
= The distance between C and the mid-point of AC
$$= \sqrt{(7-5)^2 + (9-6)^2}$$ (M1) for valid approach
$$= 3.605551275$$
$$= 3.61 \text{ km}$$ A1   N2

[4]

(c)   The post office A   A1   N1

[1]

Exercise 34

1. The diagram below shows the Voronoi diagram of three restaurants for take-away meals, A, B and C, in a town, where 1 unit represents 1 km.

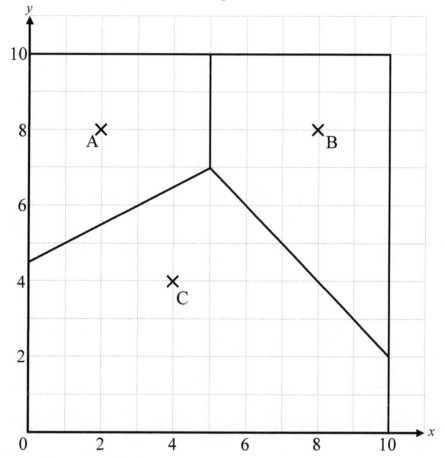

   (a) Find the area of the Voronoi cell of the restaurant B.

   [2]

   The straight line L is the boundary separating the Voronoi cells of A and C.

   (b) (i) Find the gradient of L.

   (ii) Hence, find the equation of L, giving the answer in slope-intercept form.

   [4]

   So Yeon would like to find a restaurant closest to her home to minimize the delivery time of her meal. The position of her home is at $(9, 3)$.

   (c) State the reason that she is indifferent from choosing the restaurant B and the restaurant C.

   [1]

2. The diagram below shows the Voronoi diagram of four police stations, A, B, C and D, in a city, where 1 unit represents 1 km.

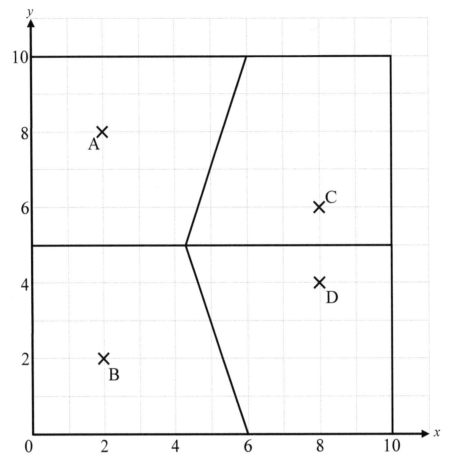

The straight line L is the boundary separating the Voronoi cells of A and C.

(a) (i) Find the gradient of L.

(ii) Hence, find the equation of L, giving the answer in slope-intercept form.

[4]

Takehiro would like to find a police station closest to his current position to report a crime. His current position is at $(5, 2)$.

(b) State which police station would be the best choice for him.

[1]

The position of James' home is at $(x, 3)$, and his home is closest to the police station D only.

(c) Write down the range of values of x.

[1]

3. The positions of three schools A, B and C in a rectangular coordinate plane are at $(4, 2)$, $(2, 4)$ and $(6, 4)$ respectively. Anna is applying a school for her son. She wants to find a school which is closest to her house. The position of her home is at $(4.15, 4)$.

    (a)   (i)   Find the distance between her home and the school A.

           (ii)   Write down the distance between her home and the school B.

           (iii)   Write down the distance between her home and the school C.

                                                                                [4]

    (b)   Write down the school in which Anna is going to apply for her son.

                                                                               [1]

4. The positions of two fire stations A and B in a rectangular coordinate plane are at $(2.5, 2.5)$ and $(7.5, 7.5)$ respectively.

    (a)   Find the coordinates of the mid-point of AB.

                                                                               [2]

A building is located at a point on the line segment AB.

    (b)   Write down the range of the x-coordinates of the position of the building if it lies in the Voronoi cell of the fire station A.

                                                                               [1]

A park is located at $(5.5, 6.5)$.

    (c)   Show that the park lies in the Voronoi cell of the fire station B.

                                                                               [3]

35 Paper 1 – Incremental Algorithm

Example

The Voronoi diagram of two points A(2, 8) and B(2, 4) are formed.

(a) Write down the equation of the boundary separating the Voronoi cells of A and B.

[1]

A new point C(8, 6) is added to this Voronoi diagram, such that the gradient of L, the boundary separating the Voronoi cells of B and C, is -3. This is shown in the following diagram:

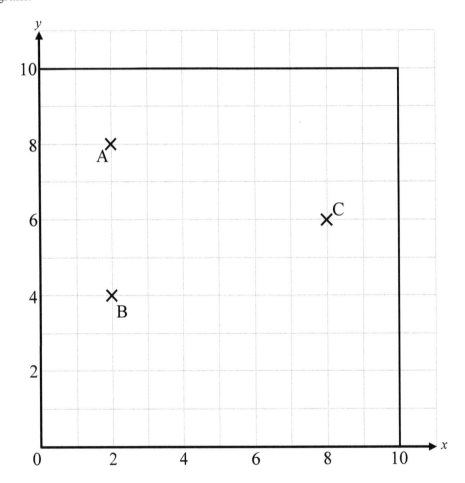

(b) (i) Write down the coordinates of the mid-point of BC.

(ii) Find the equation of L, giving the answer in slope-intercept form.

(iii) Hence, find the coordinates of the intersection of the boundaries.

[5]

Solution

(a) $y = 6$ A1   N1

[1]

(b) (i) $(5, 5)$ A1   N1

    (ii) The equation of L:
$$y - 5 = -3(x - 5)$$ (M1) for substitution
$$y = -3x + 20$$ A1   N2

    (iii) $6 = -3x + 20$ (M1) for substitution
$$3x = 14$$
$$x = \frac{14}{3}$$

Thus, the required coordinates are $\left(\frac{14}{3}, 6\right)$.   A1   N2

[5]

Your Practice Set – Applications and Interpretation for IBDP Mathematics

Exercise 35

1. The Voronoi diagram of two points P(2, 2) and Q(8, 2) are formed.

   (a) Write down the equation of the boundary separating the Voronoi cells of P and Q.

   [1]

   A new point R(6, 6) is added to this Voronoi diagram, such that the equation of L, the boundary separating the Voronoi cells of Q and R, is $x - 2y + k = 0$, where k is a constant. This is shown in the following diagram:

   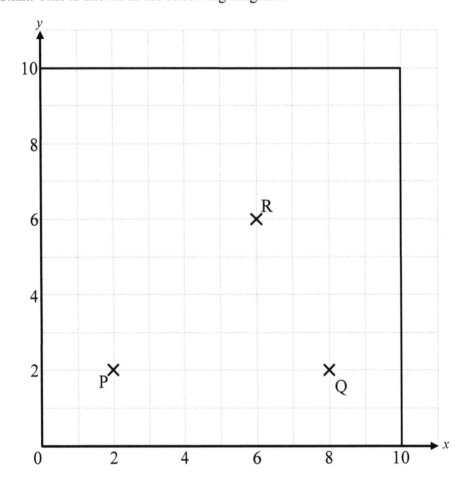

   (b) (i) Write down the coordinates of the mid-point of QR.

   (ii) Find the value of k.

   (iii) Hence, find the coordinates of the intersection of the boundaries.

   [5]

2. The positions of two car parks A and B in a rectangular coordinate plane are at $(4, 8)$ and $(8, 4)$ respectively. It is given that the boundary separating the Voronoi cells of A and B passes through the origin.

    (a) (i) Find the coordinates of the mid-point of AB.

    (ii) Hence, write down the equation of the boundary separating the Voronoi cells of A and B, giving the answer in slope-intercept form.

    [3]

    Another car park C(2, 4) is newly built, as shown in the following diagram:

    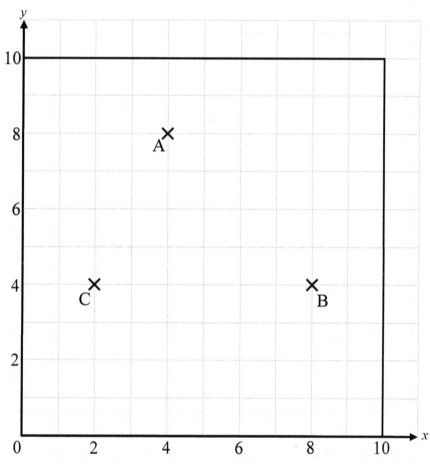

    (b) (i) Write down the equation of the boundary separating the Voronoi cells of B and C.

    (ii) Find the coordinates of the intersection of the boundaries.

    (iii) Hence, find the radius of the circle passing through all three car parks.

    [4]

Your Practice Set – Applications and Interpretation for IBDP Mathematics

3. The positions of two supermarkets A and B in a rectangular coordinate plane are at (2, 2) and (8, 8) respectively. It is given that the boundary separating the Voronoi cells of A and B has the equation $x + y = k$, where k is a constant.

   (a) (i) Find the coordinates of the mid-point of AB.

   (ii) Hence, find the value of k.

   [4]

   Another supermarket C(6, 4) is newly opened, as shown in the following diagram:

   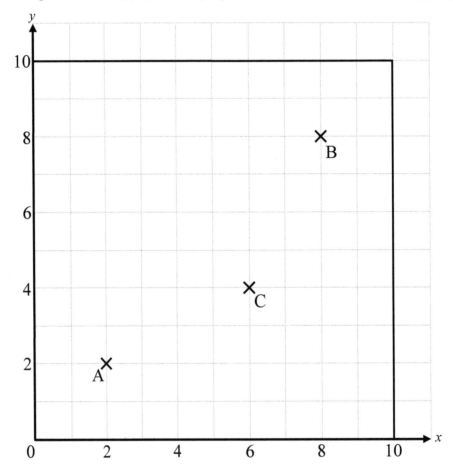

   The equation of the boundary separating the Voronoi cells of B and C is $y = -0.5x + 9.5$.

   (b) Find the coordinates of the intersection of the boundaries.

   [3]

   (c) State the significance of the Voronoi cell of B.

   [1]

4. The Voronoi diagram of three points A(1, 9), B(9, 9) and C(9, 1) are formed.

   (a) (i) Write down the equation of the boundary separating the Voronoi cells of A and B.

   (ii) Write down the equation of the boundary separating the Voronoi cells of B and C.
   [2]

   A new point D(3, 3) is added to this Voronoi diagram, such that the coordinates of the two points of intersection of the boundaries are $(5, k)$ and $(k, 5)$, where k is a constant. This is shown in the following diagram:

   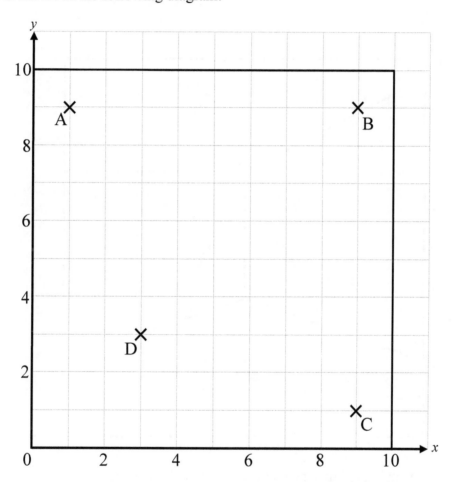

   (b) (i) Write down the coordinates of the mid-point of BD.

   (ii) Hence, find the value of k.

   (iii) Find the decrease in the area of the Voronoi cell of B after D is added.
   [6]

36 Paper 1 – Toxic Waste Dump Problems

Example

The diagram below shows the Voronoi diagram of four incinerators, A, B, C and D, in a city, where 1 unit represents 1 km. The points $E(5, 7)$ and $F\left(5, \dfrac{11}{3}\right)$ are the intersections of the boundaries of Voronoi cells.

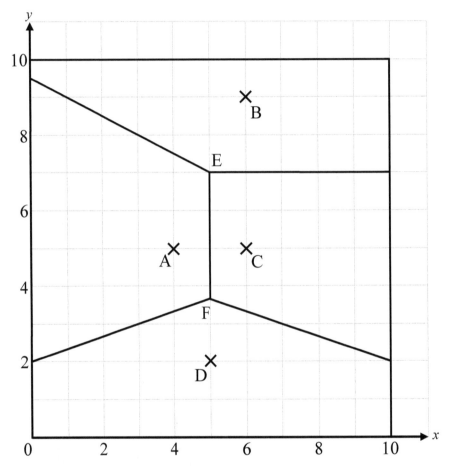

Hannah is planning to buy a new house in the town. She hopes that the location of the new house to be as far as possible from all other incinerators. There are two new houses, one is located at E and the another one is located at F.

(a) State the significance of the Voronoi cell of the incinerator C.

[1]

Two circles are constructed such that the centres are E and F respectively, and passes through the incinerators in adjacent Voronoi cells.

(b) (i) Find the radius of the circle centred at E.

(ii) Write down the exact radius of the circle centred at F.

(iii) Hence, determine which location is farthest from all four incinerators.

[4]

Solution

(a) Every position in the Voronoi cell of C has C to be the nearest incinerator.   A1   N1

[1]

(b) (i) The radius
$$= \sqrt{(6-5)^2 + (9-7)^2}$$ (M1) for valid approach
$$= 2.236067978$$
$$= 2.24 \text{ km}$$ A1   N2

(ii) $\dfrac{5}{3}$ km   A1   N1

(iii) E   A1   N1

[4]

Your Practice Set – Applications and Interpretation for IBDP Mathematics

Exercise 36

1. The diagram below shows the Voronoi diagram of four hair salons, A, B, C and D, in a city, where 1 unit represents 1 km. The points E(4.125, 5) and F(5, 5) are the intersections of the boundaries of Voronoi cells.

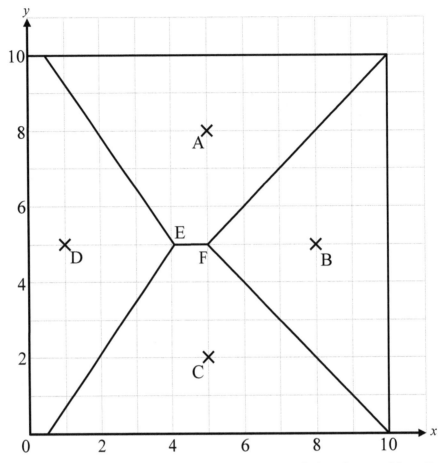

Raina is going to start her business by opening a new hair salon at either E or F. She hopes that the location of the new hair salon to be as far as possible from all other hair salons.

Two circles are constructed such that the centres are E and F respectively, and passes through the hair salons in adjacent Voronoi cells.

(a) (i) Find the exact radius of the circle centred at E.

(ii) Write down the radius of the circle centred at F.

(iii) Hence, determine which location is farthest from all four hair salons.

[4]

(b) Write down the total distance from F to all other four hair salons.

[1]

2. The diagram below shows the Voronoi diagram of four landfills, A, B, C and D, in a city, where 1 unit represents 1 km. The points E(5, 7.75) and F(5, 4.5) are the intersections of the boundaries of Voronoi cells.

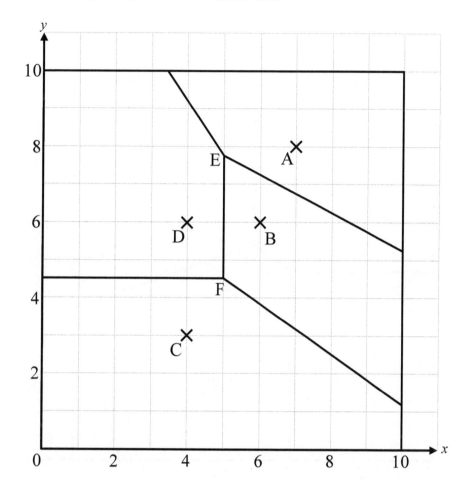

There are two hotels, one is located at E and the another one is located at F. Terence is going to book a room in a hotel. He hopes that the location of the hotel selected to be as far as possible from all other hair salons.

Two circles are constructed such that the centres are E and F respectively, and passes through the landfills in adjacent Voronoi cells.

(a) (i) Find the radius of the circle centred at E.

(ii) Find the radius of the circle centred at F.

(iii) Hence, determine which hotel is farthest from all four landfills.

[5]

(b) Find the difference of the circumferences of the two circles in (a).

[2]

Your Practice Set – Applications and Interpretation for IBDP Mathematics

3. The diagram below shows the Voronoi diagram of four farms, A, B, C and D, in a rural town, where 1 unit represents 1 km. The points $P(4, 5)$ and $Q\left(5, \dfrac{17}{3}\right)$ are the intersections of the boundaries of Voronoi cells.

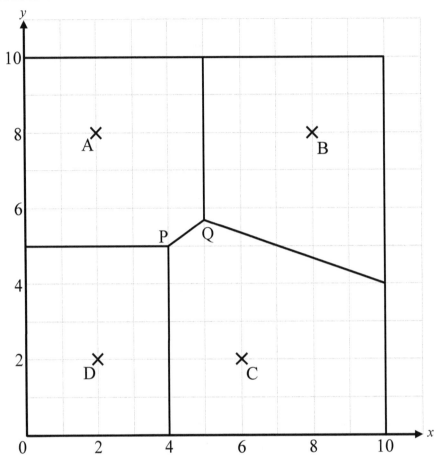

The government is planning to install a wireless network hotspot to provide internet services in the town. However, farmers in the town hope that the location of the hotspot should be farthest from all farms.

Two circles are constructed such that the centres are P and Q respectively, and passes through the farms in adjacent Voronoi cells.

(a) (i) Find the radius of the circle centred at P.

   (ii) Find the radius of the circle centred at Q.

   (iii) Hence, determine which location is farthest from all four farms.

[5]

It is given that an electronic device cannot access to the wireless network if the horizontal distance between the device and the hotspot is longer than 3.65 km.

(b) State which farm will not be able to access to the wireless network if the hotspot is installed at P.

[1]

4. The diagram below shows the Voronoi diagram of four apartments, A, B, C and D, in a city, where 1 unit represents 1 km. The points R and S$\left(\dfrac{14}{3}, \dfrac{9}{2}\right)$ are the intersections of the boundaries of Voronoi cells.

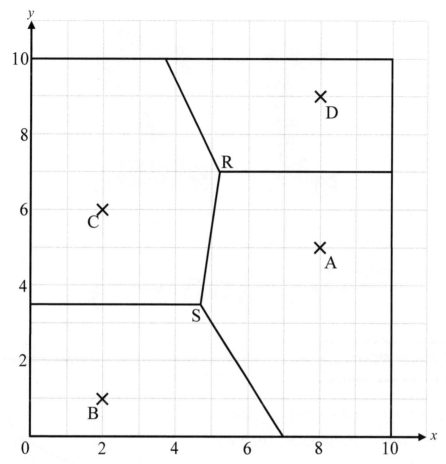

A new power station is going to be built in the city. It is suggested that the location of the power station should be farthest from all apartments.

Two circles are constructed such that the centres are R and S respectively, and passes through the apartments in adjacent Voronoi cells.

(a) Find the radius of the circle centred at S.

[2]

The circle centred at S is the largest circle such that no apartment is located in the circle. The radius of the circle centred at R is estimated as the greatest positive integer value less than the radius of the circle centred at S.

(b) Write down the estimated value of the radius of the circle centred at R.

[1]

It is given that the actual value of the radius of the circle centred at R is $\dfrac{\sqrt{185}}{4}$.

(c) Find the percentage error in this estimate.

[2]

Chapter 12

Trigonometry

SUMMARY POINTs

✓ Consider a right-angled triangle ABC:

$AB^2 + BC^2 = AC^2$: Pythagoras' Theorem

$$\begin{cases} \sin\theta = \dfrac{AB}{AC} \\ \cos\theta = \dfrac{BC}{AC} \\ \tan\theta = \dfrac{AB}{BC} \end{cases}$$: Trigonometric ratios

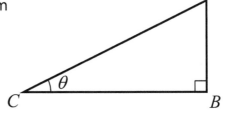

✓ Properties of a general trigonometric function $y = A\sin B(x-C) + D$:

$A = \dfrac{y_{max} - y_{min}}{2}$: Amplitude

$B = \dfrac{360°}{\text{Period}}$ and $D = \dfrac{y_{max} + y_{min}}{2}$

C can be found by substitution of a point on the graph

Solutions of Chapter 12

37 Paper 1 – Properties of Graphs

Example

The diagram below shows part of the graph of a function f.

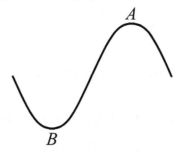

The graph has a maximum at $A(5, 4)$ and a minimum at $B(3, -2)$.

The function f can be written in the form $f(x) = p\sin(qx) + r$. Find the value of

(a) p;

[2]

(b) q;

[2]

(c) r.

[2]

Solution

(a) $p = \dfrac{4-(-2)}{2}$ (M1) for valid approach

$p = 3$ A1   N2

[2]

(b) The period of the graph is 4.

$\therefore \dfrac{360°}{q} = 4$ (M1) for valid approach

$q = 90°$ A1   N2

[2]

(c) $r = \dfrac{4+(-2)}{2}$ (M1) for valid approach

$r = 1$ A1   N2

[2]

Exercise 37

1. The diagram below shows part of the graph of a function f.

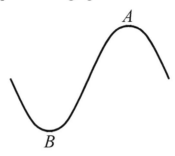

The graph has a maximum at A(45°, 2) and a minimum at B(−45°, −6).

The function f can be written in the form $f(x) = p\sin(qx) + r$. Find the value of

(a) p;
[2]

(b) q;
[2]

(c) r.
[2]

2. The diagram below shows part of the graph of a function f.

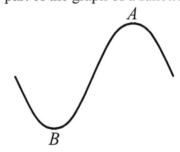

The graph has a maximum at A(3240°, 60) and a minimum at B(2520°, 28).

The function f can be written in the form $f(x) = p\sin(qx) + r$. Find the value of

(a) p;
[2]

(b) q;
[2]

(c) r.
[2]

3. The diagram below shows part of the graph of a function f.

    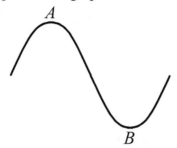

    The graph has a maximum at A(15°, 2π) and a minimum at B(45°, -2π).

    The function f can be written in the form $f(x) = p\cos(q(x-r))$, where $0° < r < 45°$. Find the value of

    (a) p;
    [2]

    (b) q;
    [2]

    (c) r.
    [2]

4. The diagram below shows part of the graph of a function f.

    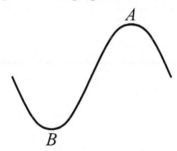

    The graph has a maximum at A(1080°, 20) and a minimum at B(540°, 0).

    The function f can be written in the form $f(x) = p\cos(q(x+r)) + 10$, where $-1440° < r < -720°$. Find the value of

    (a) p;
    [2]

    (b) q;
    [2]

    (c) r.
    [2]

Your Practice Set – Applications and Interpretation for IBDP Mathematics

Paper 1 – Sketching Trigonometric Functions

Example

Let $f(x) = 4\cos(90°x)$, for $0 \leq x \leq 4$.

(a) (i) Write down the amplitude of f.

(ii) Find the period of f.

[3]

(b) On the following grid sketch the graph of f.

[4]

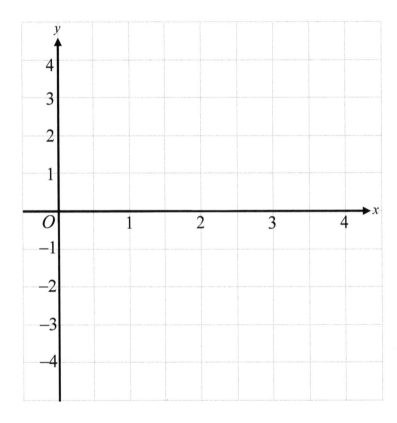

Solution

(a) (i) The amplitude of f is 4.   A1   N1

(ii) The period of f
$= 360° \div 90°$ (M1) for valid approach
$= 4$ A1   N2

[3]

(b) For correct *x*-intercepts   A1
For correct maximum and minimum points   A1
For correct domain   A1
For sinusoidal curve starting at (0, 4) and correct period   A1   N4

[4]

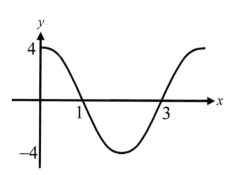

Exercise 38

1. Let $f(x) = 3.5\sin(180°x)$, for $0 \leq x \leq 4$.

   (a) (i) Write down the amplitude of f.

   (ii) Find the period of f.

   [3]

   (b) On the following grid sketch the graph of f.

   [4]

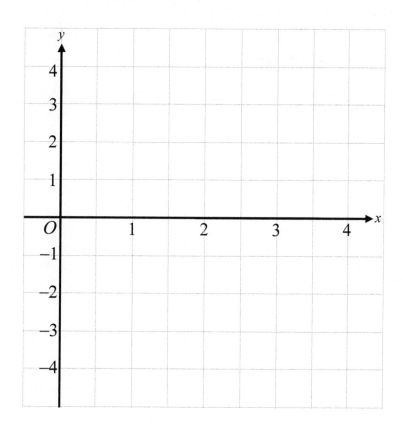

Your Practice Set – Applications and Interpretation for IBDP Mathematics

2. Let $f(x) = 3\cos(180°x) + 1$, for $-1 \leq x \leq 4$.

    (a)   (i)   Write down the amplitude of f.

           (ii)   Find the period of f.

[3]

    (b)   On the following grid sketch the graph of f.

[4]

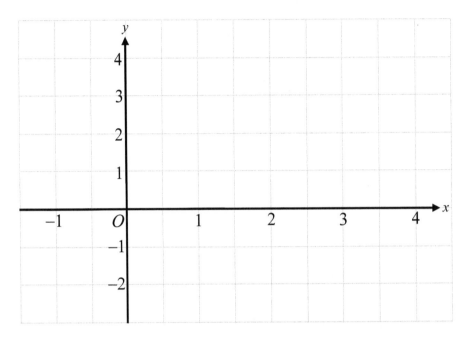

3. Let $f(x) = 4\sin(x - 90°)$, for $0° \leq x \leq 360°$.

    (a)   (i)   Write down the amplitude of f.

           (ii)   Find the period of f.

[3]

(b) On the following grid sketch the graph of f.

[4]

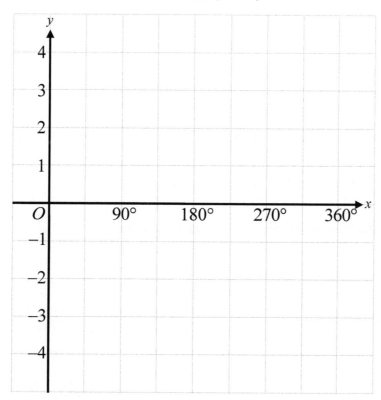

4. Let $f(x) = 3\cos\left(\dfrac{1}{2}x\right)$, for $0° \leq x \leq 360°$.

(a) (i) Write down the amplitude of f.

(ii) Find the period of f.

[3]

(b) On the following grid sketch the graph of f.

[4]

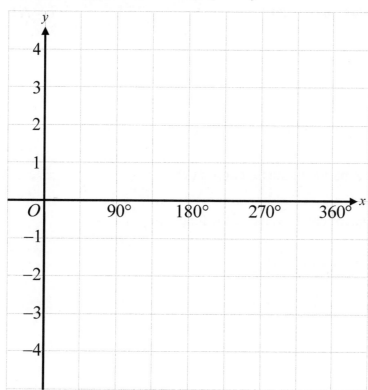

Your Practice Set – Applications and Interpretation for IBDP Mathematics

Paper 1 – Trigonometric Equations

Example

Consider the trigonometric function $f(x) = 2\cos x + 1$, where $-360° \leq x \leq 360°$.

(a) Write down the period of the function $f(x)$.

[1]

(b) Find the maximum value of the function $f(x)$.

[2]

(c) Solve the equation $f(x) = -1$.

[3]

Solution

(a)   360°   A1   N1

[1]

(b) The maximum value
$= 2(1) + 1$ (M1) for valid approach
$= 3$ A1   N2

[2]

(c) $f(x) = -1$
$2\cos x + 1 = -1$ (M1) for setting equation
$2\cos x + 2 = 0$
$x = -180°$ or $x = 180°$ A2   N3

[3]

Exercise 39

1. Consider the trigonometric function $f(x) = \sin 2x - 1$, where $0° \leq x \leq 360°$.

(a) Find the period of the function $f(x)$.

[2]

(b) Find the minimum value of the function $f(x)$.

[2]

(c) Solve the equation $f(x) = 0$.

[2]

2. Consider the trigonometric function $f(x) = 1 - 3\cos x$, where $0° \leq x \leq 720°$.

    (a) Write down the amplitude of the function $f(x)$.

[1]

    (b) Find the maximum value of the function $f(x)$.

[2]

    (c) Solve the equation $f(x) = -2$.

[3]

3. Consider the trigonometric function $f(x) = 2\sin 4x + 3$, where $-90° \leq x \leq 90°$.

    (a) Find the period of the function $f(x)$.

[2]

    (b) Write down the range of the function $f(x)$.

[2]

    (c) Solve the equation $f(x) - 4 = 0$.

[3]

4. Consider the trigonometric function $f(x) = -\dfrac{1}{2}\cos\dfrac{1}{3}x - \dfrac{1}{4}$, where $0° \leq x \leq 1080°$.

    (a) Write down the amplitude of the function $f(x)$.

[1]

    (b) Write down the range of the function $f(x)$.

[2]

    (c) Solve the equation $2f(x) + 1 = 0$.

[3]

Your Practice Set – Applications and Interpretation for IBDP Mathematics

 Paper 2 – Applications in Daily Lives

Example

At a beach the height of the water in metres is modelled by the function $h(t) = p\cos(qt) + r$, where t is the number of hours after 20:00 hours on 1 March 2018, for $0 \leq t \leq 60$.

The point $A(6, 0.8)$ represents the first low tide and $B(12, 2)$ represents the next high tide.

(a) (i) How much time is there between the first low tide and the next high tide?

    (ii) Find the difference in height between low tide and high tide.

[4]

(b) Find the value of

    (i) p;

    (ii) q;

    (iii) r.

[7]

(c) There are two low tides on 3 March 2018. At what time does the second low tide occur?

[3]

Solution

(a) (i) The amount of time
    $= 12 - 6$ (M1) for valid approach
    $= 6$ hours   A1   N2

    (ii) The difference
    $= 2 - 0.8$ (M1) for valid approach
    $= 1.2$ m   A1   N2

[4]

(b) (i) $p = 1.2 \div 2$ (M1) for valid approach
           $p = 0.6$ A1   N2

(ii)   Period $= \dfrac{360°}{q}$ (M1) for valid approach

           $q = \dfrac{360°}{12}$ (A1) for substitution

           $q = 30°$ A1   N3

(iii) $r = \dfrac{2 + 0.8}{2}$ (M1) for valid approach

           $r = 1.4$ A1   N2

                                                                               [7]

(c)   The value of t at 00:00 on 3 March 2018
     $= 4 + 24$ (M1) for valid approach
     $= 28$
     The value of t of the second low tide on 3 March 2018
     $= 30 + 12$
     $= 42$ (A1) for correct value of t
     Thus, the second low tide on 3 March 2018 occurs
     at 14:00.   A1   N3

                                                                               [3]

Exercise 40

1. In a coast the height of the sea water in metres is modelled by the function $h(t) = p\sin(qt) + r$, where t is the number of hours after 23:00 hours on 7 April 2018, for $0 \leq t \leq 72$.

   The point $A(8.25, 0.4)$ represents the first low tide and $B(13.75, 1.8)$ represents the next high tide.

   (a) (i)   How much time is there between the first low tide and the next high tide?

        (ii)   Find the difference in height between low tide and high tide.

                                                                                  [4]

(b) Find the value of

(i) p;

(ii) q;

(iii) r.

[7]

(c) There are two low tides on 9 April 2018. At what time does the second low tide occur?

[3]

2. The height of the water in metres is modelled by the function $h(t) = -p\cos(qt) + r$, where p, q and r are positive, and t is the number of hours after 12 hours on 23 August 2018, for $0 \leq t \leq 72$.

The point $A(6.5, 4.2)$ represents the first high tide and $B(13, 1.8)$ represents the next low tide.

(a) (i) How much time is there between the first high tide and the next high tide?

(ii) Find the difference in height between low tide and high tide.

[4]

(b) Find the value of

(i) p;

(ii) q;

(iii) r.

[6]

(c) Find the time when the second low tide on 25 August 2018 occur.

[3]

3. The height of a seat on a Ferris wheel in metres is modelled by the function $h(t) = -p\cos(qt) + r$, where p, q and r are positive, and t is the number of minutes after 9:00.

The seat starts at the lowest point and its height is one metre above the ground. The radius of the wheel is 45 metres and the first time the seat is 46 metres above the ground is 9:09.

(a) Find the time for the seat to be 91 metres above the ground for the first time.

[2]

(b) Find the value of

(i) p;

(ii) q;

(iii) r.

[6]

(c) Find the time for the seat to be 60 metres above the ground for the fifth time.

[4]

4. The height of a seat on a Ferris wheel in metres is modelled by the function $h(t) = -p\cos(qt) + r$, where p, q and r are positive, and t is the number of minutes after 12:45.

The seat starts at the lowest point and its height is 3 metres above the ground. The diameter of the wheel is 70 metres and the first time the seat is 73 metres above the ground is 12:58.

(a) Find the value of

(i) p;

(ii) q;

(iii) r.

[7]

(b) Find the height of the seat above the ground at 13:41.

[2]

(c) Find the time for the seat to be 10 metres above the ground for the sixth time.

[4]

Your Practice Set – Applications and Interpretation for IBDP Mathematics

Chapter

13

2-D Trigonometry

SUMMARY POINTs

✓ Consider a triangle ABC:

1. $\dfrac{\sin A}{a} = \dfrac{\sin B}{b}$ or $\dfrac{a}{\sin A} = \dfrac{b}{\sin B}$: Sine rule

2. $a^2 = b^2 + c^2 - 2bc\cos A$

    or $\cos A = \dfrac{b^2 + c^2 - a^2}{2bc}$: Cosine rule

3. $\dfrac{1}{2}ab\sin C$: Area of the triangle ABC

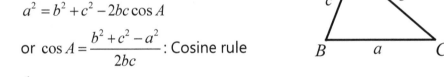

✓ Consider a sector OPQ with centre O, radius r and $\angle POQ = \theta°$:

$2\pi r \times \dfrac{\theta°}{360°}$: Arc length PRQ

$\pi r^2 \times \dfrac{\theta°}{360°}$: Area of the sector OPQ

$\pi r^2 \times \dfrac{\theta°}{360°} - \dfrac{1}{2}r^2 \sin\theta°$: Area of the segment PRQ

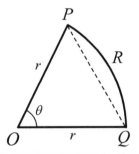

Solutions of Chapter 13

41 Paper 1 – Solving Triangles

Example

The following diagram shows triangle ABC.

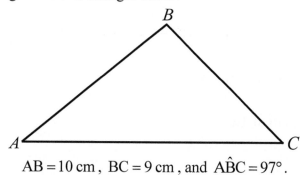

AB = 10 cm, BC = 9 cm, and $\hat{ABC} = 97°$.

(a) Find the length of AC.

[3]

(b) Find the size of the angle $\hat{BAC}$.

[3]

Solution

(a) $AC^2 = AB^2 + BC^2 - 2(AB)(BC)\cos \hat{ABC}$ (M1) for cosine rule

$AC^2 = 10^2 + 9^2 - 2(10)(9)\cos 97°$ (A1) for substitution

$AC = 14.24557762$

$AC = 14.2$ cm   A1   N3

[3]

(b) $\dfrac{\sin \hat{BAC}}{BC} = \dfrac{\sin \hat{ABC}}{AC}$ (M1) for sine rule

$\dfrac{\sin \hat{BAC}}{9} = \dfrac{\sin 97°}{14.24557762}$ (A1) for substitution

$\sin \hat{BAC} = \dfrac{9 \sin 97°}{14.24557762}$

$\hat{BAC} = 38.83397714°$

$\hat{BAC} = 38.8°$ A1   N3

[3]

Your Practice Set – Applications and Interpretation for IBDP Mathematics

Exercise 41

1. The following diagram shows triangle ABC.

    $AC = 10$ cm, $A\hat{B}C = 114°$, and $A\hat{C}B = 48°$.

    (a) Find the length of AB.
    [3]

    (b) Find the length of BC.
    [3]

2. The following diagram shows triangle ABC.

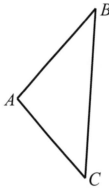

    $AC = 21$ cm, $A\hat{B}C = 43°$, and $B\hat{A}C = 92°$.

    (a) Find the length of AB.
    [3]

    (b) Find the area of the triangle ABC.
    [3]

3. The following diagram shows triangle ABC.

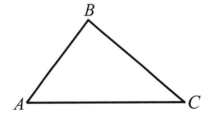

    $AC = 86$ cm, $A\hat{C}B = 40°$, and the area of triangle IBC is 1900 cm^2.

(a) Find the length of BC.

[3]

(b) Find the length of AB.

[3]

4. The following diagram shows triangle ABC.

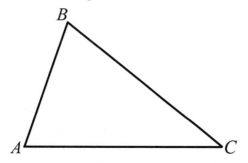

AB = 35 cm, AC = 54 cm, and the area of triangle ABC is 892 cm².

(a) Find the size of the angle BÂC.

[3]

(b) Find the length of BC.

[3]

42 Paper 1 – Angles of Elevation and Depression

Example

In the figure, the angle of elevation of the top A of a building OA from a point B is 35°. C is a point on BO such that AC = 96 m and BC = 64 m.

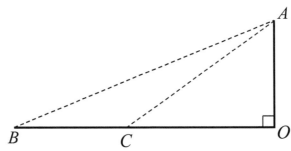

(a) Find the size of the angle BÂC.

[3]

(b) Find the size of the angle OĈA.

[2]

(c) Find the height of the building OA.

[2]

Solution

(a) $\dfrac{\sin B\hat{A}C}{64} = \dfrac{\sin 35°}{96}$ (M1)(A1) for substitution

$\sin B\hat{A}C = 0.38238429$

$B\hat{A}C = 22.48144954°$

$B\hat{A}C = 22.5°$ A1   N3

[3]

(b) $O\hat{C}A = 35° + 22.48144954°$ (M1) for valid approach

$O\hat{C}A = 57.48144954°$

$O\hat{C}A = 57.5°$ A1   N2

[2]

(c) $\sin 57.48144954° = \dfrac{OA}{96}$ (M1) for valid approach

$OA = 80.94887442$

$OA = 80.9$ m   A1   N2

[2]

Exercise 42

1. In the figure, A and B are two points on the same horizontal ground. The angles of elevation of a cable car P from A and B are 45° and 34° respectively. C is a point on AB which is vertically below P. AB = 120 m and A, B and C lie on the same vertical plane.

   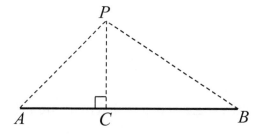

   (a) Find the size of the angle $A\hat{P}B$.

   [2]

   (b) Find the length of PB.

   [3]

   (c) Find the length of PC.

   [2]

2. In the figure, a vertical lamp post PQ stands on an inclined straight road. The inclined road makes an angle of 24° with the horizontal. R is a point on the road and 5 m away from Q. The angle of elevation of P from R is 69°.

   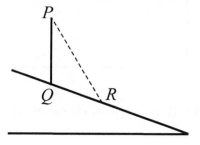

   (a) Find the size of the angle $P\hat{R}Q$.

   [2]

   (b) Find the size of the angle $Q\hat{P}R$.

   [2]

   (c) Find the length of PQ.

   [3]

3. In the figure, P and Q are two points on the opposite sides of a water well. They lie on the same horizontal level and at a vertical height of h m above a point R at the bottom of the water well. PQ = 6 m, PR = 8 m and QR = 9 m.

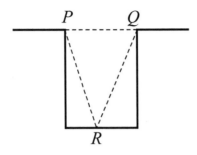

(a) Find the size of the angle of depression of R from P.

[3]

(b) Find the value of x, the horizontal distance between R and P.

[2]

(c) Find the value of h.

[2]

4. In the figure, two strings are mounted from the top F of a vertical flagpole FO to the horizontal ground at points A and B respectively. The angle of depression of A from F is 29°. A and B are 10 m apart. The length of the string BF is 30 m. A, B, F and O lie on the same vertical plane.

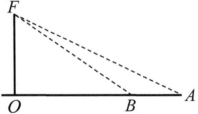

(a) Write down the size of the angle of elevation of F from A.

[1]

(b) Find the angle of elevation of F from B.

[4]

(c) Find the height of the vertical flagpole.

[2]

43 Paper 1 – Areas of Triangles and Sectors

Example

The following diagram shows a triangle ABC and a sector BDC of a circle with centre B and radius 3 cm. The points A, B and D are on the same straight line.

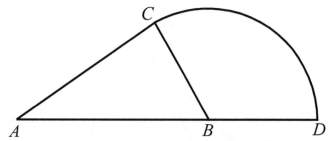

$AB = 3\sqrt{3}$ cm, $BD = 3$ cm, area of triangle $ABC = \dfrac{27}{4}$ cm², $A\hat{B}C$ is acute.

(a) Find the size of the angle $A\hat{B}C$.

[3]

(b) Hence, write down the size of the angle $C\hat{B}D$.

[1]

(c) Find the exact area of the sector CBD.

[2]

Solution

(a) Area of triangle $ABC = \dfrac{27}{4}$ cm²

$\therefore \dfrac{1}{2}(AB)(BC)\sin A\hat{B}C = \dfrac{27}{4}$ (M1) for area formula

$\because \quad D = 3$ cm

$\therefore \dfrac{1}{2}(3\sqrt{3})(3)\sin A\hat{B}C = \dfrac{27}{4}$ (A1) for substitution

$\sin A\hat{B}C = \dfrac{\sqrt{3}}{2}$

$A\hat{B}C = 60°$ A1   N3

[3]

(b) 120°   A1   N1

[1]

(c) The required area

$= \pi(3)^2 \times \dfrac{120°}{360°}$ (A1) for substitution

$= 3\pi$ cm^2 A1   N2

[2]

Exercise 43

1. The following diagram shows a triangle ABC and a sector BDC of a circle with centre B. The points A, B and D are on the same straight line.

   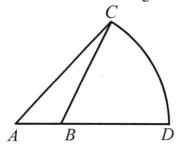

   AB = 3 cm, $A\hat{B}C = 120°$, area of triangle ABC $= \dfrac{27\sqrt{3}}{4}$ cm^2.

   (a) Write down the size of the angle $C\hat{B}D$.

   [1]

   (b) Find the length of BC.

   [3]

   (c) Find the exact arc length of the sector BDC.

   [2]

2. The following diagram shows a triangle ABC and a sector BDC of a circle with centre B and radius 10 cm. The points A, B and D are on the same straight line.

   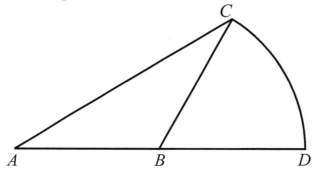

   AB = BC = 10 cm, AC = $10\sqrt{3}$ cm, $A\hat{B}C$ is obtuse.

(a) Find the size of the angle AB̂C.

[3]

(b) Hence, write down the size of the angle DB̂C.

[1]

(c) Find the perimeter of the figure ADC.

[3]

3. The following diagram shows triangle ABC, with AB = 6 cm, BC = 16 cm, and AB̂C = 60°.

(a) Show that AC = 14 cm.

[3]

(b) The shape in the following diagram is formed by adding a semicircle with diameter [AC] to the triangle.

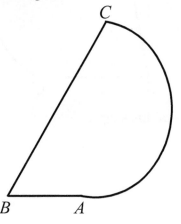

Find the area of this shape.

[3]

4. The following diagram shows triangle ABC, with BC = 8 cm, $A\hat{C}B = 60°$, and the area of the triangle ABC is $24\sqrt{3}$ cm².

(a) Show that AC = 12 cm.

[2]

(b) Find the exact length of AB.

[2]

(c) The shape in the following diagram is formed by adding a semicircle with diameter [AC] to the triangle.

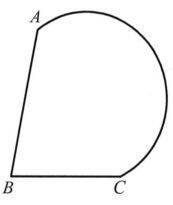

Find the perimeter of this shape.

[3]

Paper 1 – Areas and Perimeters of Sectors

Example

The following diagram shows a circle with centre O and radius 100 cm.

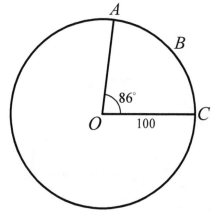

The points A, B and C are on the circumference of the circle and $A\hat{O}C = 86°$.

(a) Find the length of arc ABC.

[2]

(b) Find the perimeter of sector OABC.

[2]

(c) Find the area of sector OABC.

[2]

Solution

(a) The length of arc ABC

$= 2\pi(100) \times \dfrac{86°}{360°}$ (A1) for substitution

$= 150.0983157$

$= 150$ cm   A1   N2

[2]

(b) The perimeter of sector OABC

$= 150.0983157 + 100 + 100$ (M1) for valid approach

$= 350.0983157$

$= 350$ cm   A1   N2

[2]

(c) The area of sector OABC

$= \pi(100)^2 \times \dfrac{86°}{360°}$ (A1) for substitution

$= 7504.915784$

$= 7500 \text{ cm}^2$ A1   N2

[2]

Exercise 44

1. The following diagram shows a circle with centre O and radius 55 cm.

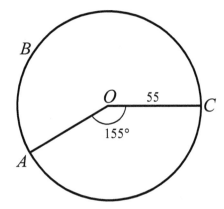

The points A, B and C are on the circumference of the circle and $A\hat{O}C = 155°$.

(a) Find the length of arc ABC.

[3]

(b) Find the perimeter of sector OABC.

[2]

(c) Find the area of sector OABC.

[2]

2. The following diagram shows a circle with centre O and radius 20 cm.

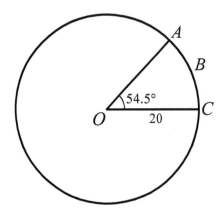

The points A, B and C are on the circumference of the circle and $A\hat{O}C = 54.5°$.

(a) Find the perimeter of sector OABC.

[4]

(b) Find the area of sector OABC.

[2]

3. The following diagram shows a circle with centre O and radius 8.6 cm.

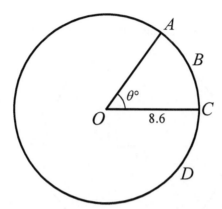

The points A, B, C and D are on the circumference of the circle such that the length of the arc ABC is 9.46 cm.

(a) Find the value of θ.

[2]

(b) Find the area of sector OADC.

[4]

4. The following diagram shows a circle with centre O.

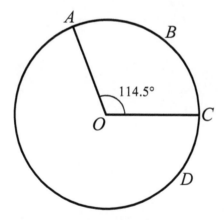

The points A, B, C and D are on the circumference of the circle such that $A\hat{O}C = 114.5°$ and the area of the sector OABC is 14 cm^2.

(a) Find the length of OC.

[2]

(b) Find the area of sector OADC.

[4]

Your Practice Set – Applications and Interpretation for IBDP Mathematics

45 Paper 1 – Areas and Perimeters of Segments

Example

The following diagram shows a circle with centre O and radius 35 cm.

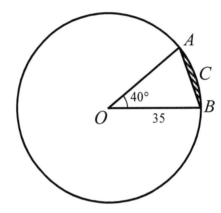

The points A, B and C are on the circumference of the circle, and $A\hat{O}B = 40°$.

(a) Find the area of the sector OACB.

[2]

(b) Find the area of the triangle OAB.

[2]

(c) Hence, find the area of the shaded segment ABC.

[2]

Solution

(a) The required area

$= \pi(35)^2 \times \dfrac{40°}{360°}$ (A1) for substitution

$= 427.6056667$

$= 428 \text{ cm}^2$ A1 N2

[2]

(b) The required area

$= \dfrac{1}{2}(35)(35)\sin 40°$ (A1) for substitution

$= 393.7074109$

$= 394 \text{ cm}^2$ A1 N2

[2]

(c) The required area
= 427.6056667 − 393.7074109   (M1) for valid approach
= 33.89825577
= 33.9 cm²   A1   N2

[2]

Exercise 45

1. The following diagram shows a circle with centre O and radius 125 cm.

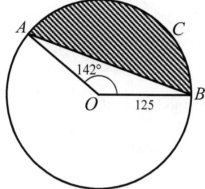

The points A, B and C are on the circumference of the circle, and $A\hat{O}B = 142°$.

(a) Find the area of the sector OACB.

[2]

(b) Find the area of the triangle OAB.

[2]

(c) Hence, find the area of the shaded segment ABC.

[2]

2. The following diagram shows a circle with centre O and radius 1740 cm.

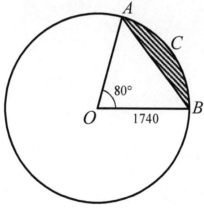

The points A, B and C are on the circumference of the circle, and $A\hat{O}B = 80°$.

(a) Find the length of arc ACB.

[2]

(b) Find the length of AB.

[3]

(c) Hence, find the perimeter of the shaded segment ABC.

[2]

3. The following diagram shows the chord AB in a circle of radius 20 cm, where AB = 32 cm.

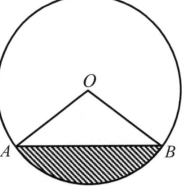

(a) Find the size of the minor angle AÔB.

[3]

(b) Find the area of the sector AOB.

[2]

It is given that the area of the triangle AOB is 192 cm^2.

(c) Find the area of the shaded region.

[2]

4. The following diagram shows the chord AB in a circle of radius 40 cm, where AB = 60 cm.

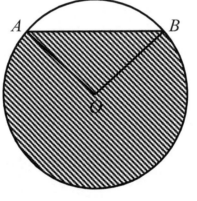

(a) Find the size of the minor angle AÔB.

[3]

(b) Find the length of the minor arc AB.

[2]

(c) Hence, find the perimeter of the shaded segment.

[2]

Paper 2 – Bearing and Speed Problems

Example

A ship is sailing south from a point A towards point D. Point C is 300 km south of A. Point D is 100 km south of C. There is a lighthouse at E. The bearing of E from A is 128°. The angle AEC is 70°. This is shown in the following diagram.

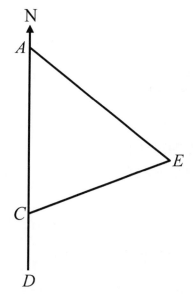

(a) Find the bearing of A from E.

[2]

(b) (i) Find the size of the angle $A\hat{C}E$.

   (ii) Hence, find the length of AE.

[5]

(c) Find the length of DE.

[3]

When the ship reaches D, it changes direction and travels directly to the lighthouse at 100 km per hour. At the same time as the ship changes direction, a ferry starts travelling to the lighthouse from a point B. This point B lies on AC, between A and C, and is the closest point to the lighthouse. The ship and the ferry arrive at the lighthouse at the same time.

(d) (i) Find the length of BE.

   (ii) Hence, find the speed of the ferry.

[5]

Your Practice Set – Applications and Interpretation for IBDP Mathematics

Solution

(a) $E\hat{A}C = 180° - 128°$
$E\hat{A}C = 52°$ (A1) for correct value
The bearing of A from E
$= 360° - 52°$
$= 308°$ A1   N2

[2]

(b) (i) $A\hat{C}E = 180° - 52° - 70°$ (M1) for valid approach
$A\hat{C}E = 58°$ A1   N2

(ii) $\dfrac{AE}{\sin 58°} = \dfrac{300}{\sin 70°}$ (M1)(A1) for substitution
$AE = 270.7421802$
$AE = 271$ km   A1   N3

[5]

(c) $DE^2 = (400)^2 + 270.7421802^2$
$ -2(400)(270.7421802)\cos 52°$ (M1)(A1) for substitution
$DE = 316.153292$
$DE = 316$ km   A1   N3

[3]

(d) (i)   Note that B lies on AC such that
$BE \perp AC$.   (M1) for valid approach
$\sin E\hat{A}C = \dfrac{BE}{AE}$
$\sin 52° = \dfrac{BE}{270.7421802}$
$BE = 213.3477495$
$BE = 213$ cm   A1   N2

(ii)   Let v km/h be the speed of the ferry.
$\dfrac{BE}{v} = \dfrac{DE}{100}$ (M1) for valid approach
$\dfrac{213.3477495}{v} = \dfrac{316.153292}{100}$ (A1) for substitution
$v = 67.4823748$
Thus, the speed of the ferry is 67.5 km/h.   A1   N3

[5]

Exercise 46

1. A ship is sailing north from a point A towards point D. Point C is 800 km north of A. Point D is 550 km north of C. There is an island at E. The bearing of E from A is 051°. The angle DCE is 77°. This is shown in the following diagram.

   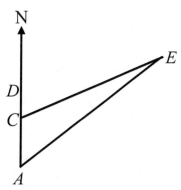

   (a) Find the bearing of C from E.

   [2]

   (b) (i) Find the size of the angle AÊC.

   (ii) Hence, find the length of AE.

   [5]

   (c) Find the length of DE.

   [3]

   (d) When the ship reaches D, it changes direction and travels directly to the island at 62 km per hour. At the same time as the ship changes direction, a boat starts travelling east, to the island from a point B. This point B is at the north of D. The ship and the boat arrive at the island at the same time.

   (i) Find the length of BE.

   (ii) Hence, find the speed of the boat.

   [5]

2. The following diagram shows the quadrilateral ABCD.

   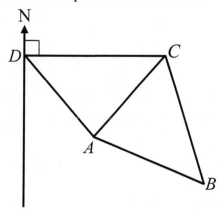

$AD = 15$ km, $A\hat{C}D = 58°$, $A\hat{C}B = 56°$, and the bearing of A from D is 160°.

(a) Find the length of AC.
[3]

(b) Find the area of triangle DAC.
[3]

The area of triangle ACD is half the area of triangle ABC.

(c) (i) Find the area of the triangle ABC

(ii) Hence, find the length of CB.
[4]

(d) A man runs from C directly to B and it takes an hour. A woman runs from D directly to B. It is given that they run at the same speed.

(i) Find the length of DC.

(ii) Find the length of BD.

(iii) Hence, find the time taken by the woman to finish her journey, giving the answer in hours.
[6]

3. The following diagram shows the quadrilateral ABCD.

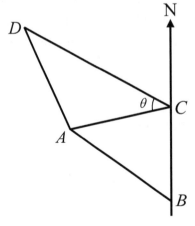

$AB = 60$ km, $DC = 83$ km, $A\hat{C}D = \theta°$, $A\hat{C}B = 83°$,
and the bearing of A from B is 312°.

(a) Find the length of AC.
[3]

(b) Find the area of triangle ABC.
[3]

The area of triangle ACD is 1.5 times the area of triangle ABC.

(c) Given that θ is acute, find θ.

[4]

(d) Car P travels from B directly to C, then from C directly to D. Car Q travels from B directly to D. The speed of Car P is 50 km per hour. Given that the time required from the two cars to finish their journey is the same, find the speed of Car Q.

[5]

4. The following diagram shows the quadrilateral ABCD.

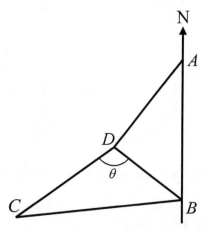

AB = 80 km, CD = 72 km, $C\hat{D}B = \theta°$, $A\hat{B}D = 61°$,
and the bearing of D from A is 220°.

(a) Find the length of DB.

[4]

(b) Find the area of triangle ABD.

[3]

The area of triangle ABD is equal to the area of triangle BCD.

(c) Given that θ is obtuse, find θ.

[4]

(d) A taxi travels from C directly to B, then from B directly to A. Given that the taxi travels at a maximum speed of 70 km per hour, find the minimum time required for the taxi to finish the journey, giving the answer to the nearest minute.

[5]

Your Practice Set – Applications and Interpretation for IBDP Mathematics

Paper 2 – Problems in Circles and Sectors

Example

Consider a circle with centre O and radius 12 cm. Triangle ABC is drawn such that its vertices are on the circumference of the circle.

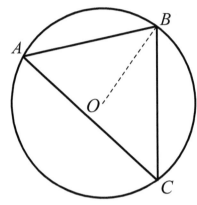

AB = 16.8 cm, BC = 21.5 cm and $B\hat{A}C = 64°$

(a) Find the size of the angle $A\hat{C}B$.

[3]

(b) (i) Find the size of the angle $A\hat{B}C$.

(ii) Hence, find the length of AC.

[5]

(c) (i) Find the size of the angle $A\hat{O}C$.

(ii) Hence, find the length of arc ABC.

[6]

Solution

(a) $\dfrac{\sin A\hat{C}B}{16.8} = \dfrac{\sin 64°}{21.5}$ (M1)(A1) for substitution

$\sin A\hat{C}B = 0.702313487$

$A\hat{C}B = 44.61291137°$

$A\hat{C}B = 44.6°$ A1   N3

[3]

(b) (i) $A\hat{B}C = 180° - 44.61291137° - 64°$ (M1) for valid approach

$A\hat{B}C = 71.38708863°$

$A\hat{B}C = 71.4°$ A1   N2

(ii) $AC^2 = 16.8^2 + 21.5^2$
$-2(16.8)(21.5)\cos 71.38708863°$ (M1)(A1) for substitution

$AC = 22.66979298$

$AC = 22.7$ cm   A1   N3

[5]

(c) (i) $\cos A\hat{O}C = \dfrac{12^2 + 12^2 - 22.66979298^2}{2(12)(12)}$ (M1)(A1) for substitution

$A\hat{O}C = 141.6691693°$

$A\hat{O}C = 142°$ A1   N3

(ii)   reflex $A\hat{O}C = 360° - 141.6691693°$ (M1) for valid approach

reflex $A\hat{O}C = 218.3308307°$

The length of arc ABC

$= 2\pi(12) \times \dfrac{218.3308307°}{360°}$ (A1) for substitution

$= 45.72710224$

$= 45.7$ cm   A1   N3

[6]

Exercise 47

1. Consider a circle with centre O and radius 14 cm. Triangle ABC is drawn such that its vertices are on the circumference of the circle.

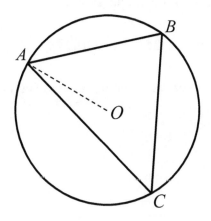

$AB = 20.8$ cm, $AC = 26.6$ cm and $A\hat{B}C = 71.5°$

(a) Find the size of the angle $A\hat{C}B$.

[3]

(b) (i) Find the size of the angle $B\hat{A}C$.

   (ii) Find the length of BC.

[5]

(c) (i) Find the size of the reflex angle $B\hat{O}C$.

   (ii) Hence or otherwise, find the area of sector OBAC.

[6]

2. Consider a circle with centre O and radius 23 cm. Triangle ABC is drawn such that its vertices are on the circumference of the circle. D is another point on the circumference.

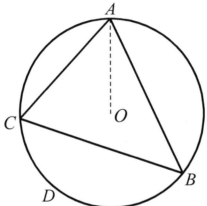

AB = 43.2 cm, $A\hat{B}C = 50°$ and $A\hat{C}B = 70°$

(a) Find the length of AC.

[3]

(b) Find the length of BC.

[5]

(c) (i) Find the size of the angle $B\hat{O}C$.

   (ii) Find the area of the sector OBDC.

   (iii) Hence or otherwise, find the area of segment BDC.

[8]

3. The following diagram shows a circle with centre O and radius 11 cm.

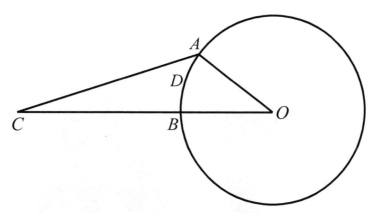

The points A, B and D lie on the circle. The point C is outside the circle, on OC. Angle ACB = 21° and angle AOB = 40°.

(a) Find the length of AC.

[3]

(b) Find the length of OC.

[4]

(c) Find the area of region ADBC.

[6]

4. The following diagram shows a circle with centre O and radius 28 cm.

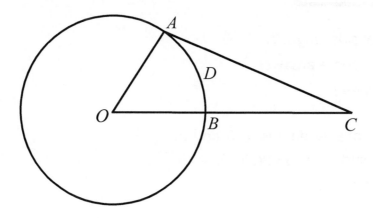

The points A, B and D lie on the circle. The point C is outside the circle, on OC. AC = 47 cm and angle AOB = 63°.

(a) Find the size of the angle AĈB.

[3]

(b) Find the length of OC.

[4]

(c) Find the perimeter of region ADBC.

[4]

Your Practice Set – Applications and Interpretation for IBDP Mathematics

Chapter

14

Areas and Volumes

SUMMARY POINTs

- ✓ For a cube of side length l:
    1. Total surface area $= 6l^2$
    2. Volume $= l^3$

- ✓ For a cuboid of side lengths a, b and c:
    1. Total surface area $= 2(ab + bc + ac)$
    2. Volume $= abc$

- ✓ For a prism of height h and cross-sectional area A:
    1. Volume $= Ah$

- ✓ For a cylinder of height h and radius r:
    1. Total surface area $= 2\pi r^2 + 2\pi rh$
    2. Lateral surface area $= 2\pi rh$
    3. Volume $= \pi r^2 h$

SUMMARY POINTs

- For a pyramid of height h and base area A:
    1. Volume $= \dfrac{1}{3} A h$

- For a circular cone of height h and radius r:
    1. Slant height $l = \sqrt{r^2 + h^2}$
    2. Total surface area $= \pi r^2 + \pi r l$
    3. Curved surface area $= \pi r l$
    4. Volume $= \dfrac{1}{3} \pi r^2 h$

- For a sphere of radius r:
    1. Total surface area $= 4\pi r^2$
    2. Volume $= \dfrac{4}{3} \pi r^3$

- For a hemisphere of radius r:
    1. Total surface area $= 3\pi r^2$
    2. Curved surface area $= 2\pi r^2$
    3. Volume $= \dfrac{2}{3} \pi r^3$

 Solutions of Chapter 14

Your Practice Set – Applications and Interpretation for IBDP Mathematics

Paper 1 – Formulae of Areas and Volumes

Example

The area of a circle is equal to 12 cm^2.

(a) Find the radius of the circle.

[2]

This circle is the base of a **solid** cylinder of height 20 cm.

(b) Write down the volume of the **solid** cylinder.

[1]

(c) Find the **total** surface area of the **solid** cylinder.

[3]

Solution

(a) $\pi r^2 = 12$ (M1) for setting equation

$r = \sqrt{\dfrac{12}{\pi}}$

$r = 1.954410048$

$r = 1.95$ cm A1 N2

[2]

(b) 240 cm^3 A1 N1

[1]

(c) The total surface area

$= 2\pi r^2 + 2\pi rh$ (M1) for valid approach

$= 2\pi(1.954410048)^2 + 2\pi(1.954410048)(20)$ (A1) for substitution

$= 269.59841$

$= 270 \text{ cm}^2$ A1 N3

[3]

Exercise 48

1. The area of a circle is equal to 9π cm^2.

    (a) Find the radius of the circle.
    [2]

    This circle is the base of a **solid** circular cone of height 4 cm.

    (b) Write down the volume of the **solid** circular cone, giving the answer in terms of π.
    [1]

    (c) Find the **total** surface area of the **solid** circular cone.
    [4]

2. The area of a circle is equal to 37 cm^2.

    (a) Find the radius of the circle.
    [2]

    This circle is the base of a **solid** hemisphere.

    (b) Write down the volume of the **solid** hemisphere.
    [2]

    (c) Find the **total** surface area of the **solid** hemisphere.
    [3]

3. A type of biscuit is packaged in a right circular cone that has volume 150 cm^3 and vertical height 13 cm.

    (a) Find the radius, r, of the circular base of the cone.
    [2]

    (b) Find the slant height, l, of the cone.
    [2]

    (c) Find the curved surface area of the cone.
    [2]

4. Chocolate ice cream is designed in a right circular cone that has curved surface area 369π cm^2 and slant height 41 cm.

    (a) Find the radius, r, of the circular base of the cone.
    [2]

    (b) Find the vertical height, h, of the cone.
    [2]

    (c) Find the volume of the cone, giving the answer in terms of π.
    [2]

Your Practice Set – Applications and Interpretation for IBDP Mathematics

Paper 1 – Composite Objects

Example

A museum is built in the shape of a cylinder with a hemispherical roof on the top. The height of the cylinder is 18 m and its radius is 24 m.

(a) Calculate the volume of the museum.

[4]

The hemispherical roof and the outer cylindrical wall are to be painted.

(b) Calculate the area to be painted.

[3]

Solution

(a) The volume

$= \dfrac{2}{3} \pi r^3 + \pi r^2 h$ (M2) for valid approach

$= \dfrac{2}{3} \pi (24)^3 + \pi (24)^2 (18)$ (A1) for substitution

$= 61524.95053$

$= 61500 \text{ m}^3$ A1   N4

[4]

(b) The area

$= 2\pi r^2 + 2\pi r h$ (M1) for valid approach

$= 2\pi (24)^2 + 2\pi (24)(18)$ (A1) for substitution

$= 6333.45079$

$= 6330 \text{ m}^2$ A1   N3

[3]

Exercise 49

1. A church is built in the shape of a cylinder with a roof in the shape of a circular cone on the top. The height of the cylinder is 5 m and its radius is 12 m. The height of the circular cone is 16 m.

    (a) Calculate the volume of the church.

    [4]

    The roof on the top of the church is to be painted.

    (b) Calculate the area that is to be painted.

    [3]

2. A kid's plastic toy consists of a hemisphere attached to a circular cone with the same base radius. The height and the radius of the cone are 6 cm and 8 cm respectively.

    (a) Calculate the volume of the toy.

    [4]

    (b) Calculate the total surface area of the toy.

    [3]

3. A building is built in the shape of a cylinder with a hemispherical roof on the top. The height of the cylinder is 40 m and the volume of the building is 54000π m^3.

    (a) Calculate the radius of the cylinder.

    [5]

    The hemispherical roof is to be painted.

    (b) Calculate the area that is to be painted.

    [2]

4. A tablet produced has a shape of a cylinder with an identical hemisphere at each end of the cylinder. The height of the cylinder is 3 mm and the total surface area of the tablet is 28π mm^2.

    (a) Calculate the radius of the cylinder.

    [5]

    (b) Calculate the volume of the tablet.

    [2]

Your Practice Set – Applications and Interpretation for IBDP Mathematics

Paper 1 – Remoulding Problems

Example

A metal sphere has a radius 8.8 cm.

(a) Find the volume of the sphere expressing your answer in the form $a \times 10^k$, $1 \leq a < 10$ and $k \in \mathbb{Z}$.

[3]

The sphere is to be melted down and remoulded into the shape of a cone with a height of 17 cm.

(b) Find the radius of the base of the cone.

[3]

Solution

(a) The volume
$$= \frac{4}{3}\pi r^3$$ (M1) for valid approach
$$= \frac{4}{3}\pi (8.8)^3$$
$$= 2854.543238$$ (A1) for correct value
$$= 2850$$
$$= 2.85 \times 10^3 \text{ cm}^3$$ A1 N3

[3]

(b) $V = \frac{1}{3}\pi r^2 h$ (M1) for valid approach

$2854.543238 = \frac{1}{3}\pi r^2 (17)$ (A1) for substitution

$r^2 = 160.3463529$
$r = 12.66279404$
$r = 12.7$ cm A1 N3

[3]

Exercise 50

1. A metal hemisphere has a radius 22 cm.

   (a) Find the volume of the hemisphere expressing your answer in the form $a \times 10^k$, $1 \leq a < 10$ and $k \in \mathbb{Z}$.

   [3]

   The hemisphere is to be melted down and remoulded into the shape of a cylinder with a height of 26 cm.

   (b) Find the radius of the base of the cylinder.

   [3]

2. A metal square-based pyramid has a height 35 cm and a base length 8π cm.

   (a) Find the volume of the pyramid expressing your answer in the form $a \times 10^k$, $1 \leq a < 10$ and $k \in \mathbb{Z}$.

   [3]

   The pyramid is to be melted down and remoulded into the shape of a sphere.

   (b) Find the radius of the sphere.

   [3]

3. A metal cylinder has a height 100 cm and a radius 7 cm.

   (a) Find the volume of the cylinder expressing your answer in terms of π.

   [2]

   The cylinder is to be melted down and remoulded into ten identical metal hemispheres.

   (b) Find the radius of one hemisphere expressing your answer in the form $a \times 10^k$, $1 \leq a < 10$ and $k \in \mathbb{Z}$.

   [4]

4. The radius and the vertical height of a metal circular cone are both equal to 27 cm.

   (a) Find the volume of the circular cone expressing your answer in the form $a \times 10^k$, $1 \leq a < 10$ and $k \in \mathbb{Z}$.

   [3]

   4 identical circular cones are to be melted down and remoulded into 27 identical metal spheres.

   (b) Find the ratio of the radius of the cone to the radius of the sphere.

   [4]

Your Practice Set – Applications and Interpretation for IBDP Mathematics

 Paper 1 – Percentage Problems

Example

A balloon in the shape of a sphere is filled with gas until the radius is 10 cm.

(a) Calculate the volume of the balloon.

[2]

The volume of the balloon is increased by 25%.

(b) Calculate the radius of the balloon following this increase.

[4]

Solution

(a) The volume

$= \dfrac{4}{3}\pi r^3$

$= \dfrac{4}{3}\pi (10)^3$ (A1) for substitution

$= 4188.790205$

$= 4190 \text{ cm}^3$ A1   N2

[2]

(b) $V = \dfrac{4}{3}\pi r^3$

$4188.790205 \times (1 + 25\%) = \dfrac{4}{3}\pi r^3$ (M2) for valid approach

$r^3 = 1250$ (M1) for finding r^3

$r = 10.77217345$

$r = 10.8 \text{ cm}$ A1   N4

[4]

Exercise 51

1. A plastic toy is in the shape of a sphere with radius 15 cm.

   (a) Calculate the total surface area of the plastic toy.
   [2]

   The plastic toy is enlarged such that its total surface area is increased by 30%.

   (b) Calculate the radius of the plastic toy following this increase.
   [4]

2. A football in the shape of a sphere is filled with gas until the radius is 14 cm.

   (a) Calculate the total surface area of the football.
   [2]

   The total surface area of the football is increased by 15%.

   (b) Calculate the percentage increase of its volume.
   [5]

3. The radius and the height of a cylindrical sponge cake are 18 cm and 8 cm respectively.

   (a) Calculate the total surface area of the sponge cake.
   [2]

   The sponge cake is then cut into two identical halves vertically by a knife.

   (b) Calculate the percentage increase of the total surface area.
   [4]

4. The radius and the slant height of a crystal in the shape of circular cone are 7 cm and 25 cm respectively.

   (a) Calculate the total surface area of the crystal.
   [2]

   The crystal is then cut into two identical halves vertically.

   (b) Calculate the percentage increase of the total surface area.
   [5]

52 Paper 2 – Angles between Lines and Planes

Example

A solid metal cylinder has a base radius of 30 cm and a height of 45 cm.

(a) Find the volume of the cylinder.

[3]

(b) Find the total surface area of the cylinder.

[3]

The cylinder was melted and recast into a solid circular cone, shown in the following diagram. The base radius OA is 40 cm.

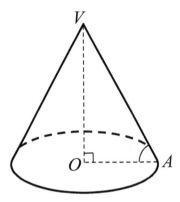

(c) Find the exact value of the height, OV, of the cone.

[3]

(d) Find the size of angle VAO.

[2]

(e) Find the total surface area of the cone.

[4]

Solution

(a) The volume
$= \pi r^2 h$ (M1) for valid approach
$= \pi (30)^2 (45)$ (A1) for substitution
$= 127234.5025$
$= 127000 \text{ cm}^3$ A1   N3

[3]

(b) The total surface area
$= 2\pi r^2 + 2\pi rh$ (M1) for valid approach
$= 2\pi(30)^2 + 2\pi(30)(45)$ (A1) for substitution
$= 14137.16694$
$= 14100 \text{ cm}^2$ A1 N3

[3]

(c) $V = \dfrac{1}{3}\pi r^2 h$ (M1) for valid approach

$127234.5025 = \dfrac{1}{3}\pi(40)^2(\text{OV})$ (A1) for substitution

$\text{OV} = 75.9375 \text{ cm}$ A1 N3

[3]

(d) $\tan \hat{\text{VAO}} = \dfrac{\text{OV}}{\text{OA}}$ (M1) for tangent ratio

$\tan \hat{\text{VAO}} = \dfrac{75.9375}{40}$

$\hat{\text{VAO}} = 62.22202722°$
$\hat{\text{VAO}} = 62.2°$ A1 N2

[2]

(e) The slant height l
$= \sqrt{75.9375^2 + 40^2}$ (M1) for finding slant height
$= 85.82833976$
The total surface area
$= \pi r^2 + \pi rl$ (M1) for valid approach
$= \pi(40)^2 + \pi(40)(85.82833976)$ (A1) for substitution
$= 15812.05551$
$= 15800 \text{ cm}^2$ A1 N4

[4]

Your Practice Set – Applications and Interpretation for IBDP Mathematics

Exercise 52

1. A solid metal hemisphere has a base radius of 10 cm.

   (a) Find the volume of the hemisphere.

   [3]

   (b) Find the total surface area of the hemisphere.

   [3]

   Four identical hemispheres were melted and recast into a solid circular cone, shown in the following diagram. The base diameter AB is 38 cm.

   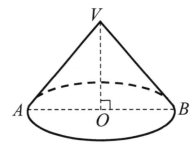

   (c) Find the value of the height, OV, of the cone.

   [3]

   (d) Find the size of angle AVB.

   [4]

   (e) Find the total surface area of the cone.

   [3]

2. A solid pentagonal prism has a base area of 260 cm^2 and a height of 100 cm. The perimeter of the pentagonal base is 62 cm.

   (a) Find the volume of the prism.

   [2]

   (b) Find the total surface area of the prism.

   [3]

   The prism was melted and recast into a solid square pyramid VABCD, shown in the following diagram. O is the centre of the base. The vertical height OV is 40 cm. M is the mid-point of AD.

   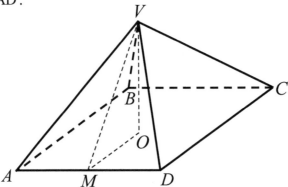

(c) Find the length of AD.

[3]

(d) Find the size of angle VMO.

[3]

(e) Find the size of angle VAO.

[5]

3. A circular cone has a base radius of 40 cm and a slant height of 104 cm, as shown in the following diagram.

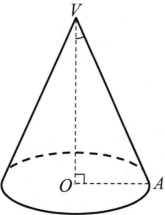

(a) Find the size of angle OVA.

[2]

(b) Find the total surface area of the cone.

[3]

(c) Find the volume of the cone.

[4]

The cone is cut into two parts horizontally such that the two parts have the same volume. The upper part is a smaller cone and the lower part is a frustum. Let R and H be the base radius and the vertical height of the upper part respectively.

(d) Express R in terms of H.

[2]

(e) Find the value of H and of R.

[6]

4. A square pyramid VABCD has a vertical height VO of 56 cm and a slant height VA of 70 cm, where O is the centre of the base, shown in the following diagram.

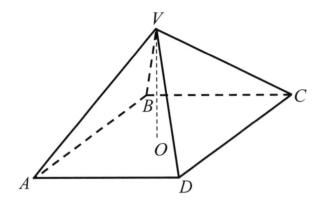

(a) Find the size of angle OVA. [2]

(b) Find the length of AD. [4]

(c) Find the volume of the cone. [3]

The pyramid is cut into two parts horizontally such that the two parts have the same volumes. The upper part is a smaller pyramid and the lower part is a frustum. Let x cm² and y cm² be the base area of the upper part and the area of the square ABCD respectively. It is given that the vertical height of the upper part is $56 \times 2^{-\frac{1}{3}}$ cm.

(d) Find $x:y$ giving the answer in the form $1:n$. [6]

Chapter 15

Differentiation

> **SUMMARY POINTs**
>
> ✓ $\dfrac{dy}{dx} = f'(x)$: Derivative of the function $y = f(x)$ (First derivative)
>
> ✓   Rules of differentiation:
> 1. $f(x) = x^n \Rightarrow f'(x) = nx^{n-1}$
> 2. $f(x) = p(x) + q(x) \Rightarrow f'(x) = p'(x) + q'(x)$
> 3. $f(x) = cp(x) \Rightarrow f'(x) = cp'(x)$
>
> ✓   Relationship between graph properties and the derivatives:
> 1. $f'(x) > 0$ for $a \leq x \leq b$: $f(x)$ is increasing in the interval
> 2. $f'(x) < 0$ for $a \leq x \leq b$: $f(x)$ is decreasing in the interval
> 3. $f'(a) = 0$: $(a, f(a))$ is a stationary point of $f(x)$
> 4. $f'(a) = 0$ and $f'(x)$ changes from positive to negative at $x = a$: $(a, f(a))$ is a maximum point of $f(x)$
> 5. $f'(a) = 0$ and $f'(x)$ changes from negative to positive at $x = a$: $(a, f(a))$ is a minimum point of $f(x)$

Your Practice Set – Applications and Interpretation for IBDP Mathematics

SUMMARY POINTs

✓ Tangents and normals:
1. $f'(a)$: Slope of tangent at $x = a$
2. $\dfrac{-1}{f'(a)}$: Slope of normal at $x = a$
3. $y - f(a) = f'(a)(x - a)$: Equation of tangent at $x = a$
4. $y - f(a) = \left(\dfrac{-1}{f'(a)}\right)(x - a)$: Equation of normal at $x = a$

Solutions of Chapter 15

53 Paper 1 – Gradients of Tangents and Normals

Example

Consider the function $f(x) = 2x^2 + 3x - 4$.

(a) Find the expression of $f'(x)$.

[2]

(b) Find the gradient of the graph of $f(x)$ at $x = 5$.

[2]

(c) Find the gradient of the normal to $f(x)$ at $x = -5$.

[2]

Solution

(a) $f'(x) = 2(2x) + 3(1) - 0$ (A1) for correct derivatives

$f'(x) = 4x + 3$ A1   N2

[2]

(b) The gradient

$= 4(5) + 3$ (A1) for substitution

$= 23$ A1   N2

[2]

(c) The gradient of the normal

$= \dfrac{-1}{4(-5) + 3}$ (A1) for substitution

$= \dfrac{1}{17}$ A1   N2

[2]

Exercise 53

1. Consider the function $f(x) = 3x^4 + \dfrac{1}{2}x$.

   (a) Find the expression of $f'(x)$.

   [2]

   (b) Find the gradient of the graph of $f(x)$ at $x = 0$.

   [2]

   (c) Find the gradient of the normal to $f(x)$ at $x = -1$.

   [2]

Your Practice Set – Applications and Interpretation for IBDP Mathematics

2. Consider the function $f(x) = 4x^2 - 6$.

    (a) Find the expression of $f'(x)$.

    [2]

    (b) Find the gradient of the graph of $f(x)$ at $x = \dfrac{1}{4}$.

    [2]

    (c) The normal to $f(x)$ at $x = a$ has gradient $\dfrac{1}{16}$. Find the value of a.

    [3]

3. Consider the function $f(x) = \dfrac{1}{2}x^2 - \dfrac{1}{x}$.

    (a) Find the value of $f(-2)$.

    [2]

    (b) Find the expression of $f'(x)$.

    [2]

    (c) The normal to $f(x)$ at $x = a$ has gradient $-\dfrac{9}{28}$, where $a > 1$. Find the value of a.

    [4]

4. Consider the function $f(x) = \dfrac{x^3}{a} + x$, where a is a positive constant.

    (a) Write down the value of the y-intercept.

    [1]

    (b) Find the expression of $f'(x)$ in terms of a.

    [2]

    (c) The normal to $f(x)$ at $x = 2$ has gradient $-\dfrac{1}{3}$. Find the value of a.

    [3]

Paper 1 – Equations of Tangents and Normals

Example

Let $f(x) = 2 + x - \dfrac{3}{x}$. The line L is the tangent to the curve of f at $(1, 0)$.

(a) Find the expression of $f'(x)$.

[2]

(b) Find the gradient of L.

[2]

(c) Find the equation of L in the form $y = ax + b$.

[2]

Solution

(a) $f'(x) = 0 + 1 - 3(-x^{-2})$ (A1) for correct derivatives

$f'(x) = 1 + 3x^{-2}$ A1   N2

[2]

(b) The gradient of L

$= 1 + 3(1)^{-2}$ (A1) for substitution

$= 4$ A1   N2

[2]

(c) The equation of L:

$y - 0 = 4(x - 1)$ (A1) for substitution

$y = 4x - 4$ A1   N2

[2]

Exercise 54

1. Let $f(x) = 6x^4 - 21x^2$. The line L is the tangent to the curve of f at $(2, 12)$.

(a) Find the expression of $f'(x)$.

[2]

(b) Find the gradient of L.

[2]

(c) Find the equation of L in the form $y = ax + b$.

[2]

Your Practice Set – Applications and Interpretation for IBDP Mathematics

2. Let $f(x) = 3x - \dfrac{4}{x^2}$. The line L is the normal to the curve of f at $(1, -1)$.

   (a) Find the expression of $f'(x)$.

   [2]

   (b) Find the gradient of L.

   [2]

   (c) Find the equation of L in the form $y = ax + b$.

   [2]

3. Let $f(x) = ax^3 - 2x^2 + 1$. The line L is the tangent to the curve of f at $(3, 27a - 17)$.

   (a) Find the expression of $f'(x)$ in terms of a.

   [2]

   (b) The gradient of L is 96. Find the value of a.

   [2]

   (c) Find the equation of L in the form $y = mx + c$.

   [2]

4. Let $f(x) = 3 - ax^3$. The line L is the normal to the curve of f at $(2, 3 - 8a)$.

   (a) Find the expression of $f'(x)$ in terms of a.

   [2]

   (b) Find the gradient of L in terms of a.

   [2]

   (c) Write down the value of a if the gradient of L is $-\dfrac{1}{12}$.

   [1]

   (d) Find the equation of L in the form $y = mx + c$.

   [2]

Paper 1 – Parallel Tangents

Example

Consider the function $f(x) = 4x^2 + 3$ and $g(x) = 2x - 5$.

(a) Find the expression of $f'(x)$.

[2]

(b) Write down the expression of $g'(x)$.

[1]

(c) Find the value of x such that the gradients of $f(x)$ and $g(x)$ are the same.

[2]

(d) Write down the value of $f(x)$ when the gradients of $f(x)$ and $g(x)$ are the same.

[1]

Solution

(a) $f'(x) = 4(2x) + 0$ (A1) for correct derivatives
$f'(x) = 8x$ A1   N2

[2]

(b) $g'(x) = 2$ A1   N1

[1]

(c) $f'(x) = g'(x)$
$8x = 2$ (M1) for setting equation
$x = \dfrac{1}{4}$ A1   N2

[2]

(d) $\dfrac{13}{4}$ A1   N1

[1]

Exercise 55

1. Consider the function $f(x) = 2x^2 - x$ and $g(x) = x^2$.

(a) Find the expression of $f'(x)$.

[2]

(b) Write down the expression of $g'(x)$.

[1]

(c) Find the value of x such that the gradients of $f(x)$ and $g(x)$ are the same.

[2]

(d) Write down the value of $g'(x)$ when the gradients of $f(x)$ and $g(x)$ are the same.

[1]

2. Consider the function $f(x) = x^3 + x^2 + 10$ and $g(x) = 5 + x$.

(a) Find the expression of $f'(x)$.

[2]

(b) Write down the expression of $g'(x)$.

[1]

(c) Find the values of x such that the gradients of $f(x)$ and $g(x)$ are the same.

[3]

(d) Write down the x-coordinate of the mid-point of AB, where A and B are the points on $f(x)$ such that the gradients of $f(x)$ and $g(x)$ are the same.

[1]

3. Consider the function $f(x) = 24x - x^3$ and $g(x) = x^3$.

(a) Find the expression of $f'(x)$.

[2]

(b) Write down the expression of $g'(x)$.

[1]

(c) Find the values of x such that the gradients of $f(x)$ and $g(x)$ are the same.

[3]

(d) A and B are the points on $g(x)$ such that the gradients of $f(x)$ and $g(x)$ are the same. Find the gradient of AB.

[2]

4. Consider the function $f(x) = 5 - ax^2$ and $g(x) = -4x$, where $a > 0$.

(a) (i) Write down the expression of $f'(x)$ in terms of a.

   (ii) Write down the expression of $g'(x)$.

[2]

(b) The gradients of $f(x)$ and $g(x)$ are the same at $x = 2$. Find the value of a.

[2]

(c) Write down the value of $f(2)$.

[1]

(d) Find the x-intercept of the tangent to $f(x)$ at $x = 2$.

[2]

56 Paper 1 – Increasing and Decreasing Functions

Example

Let $f(x) = x^3 - 48x + 102$.

(a) Write down the y-intercept of $f(x)$.

[1]

(b) Find the expression of $f'(x)$.

[2]

(c) Find the intervals of x such that $f(x)$ is increasing.

[3]

Solution

(a) 102   A1   N1

[1]

(b) $f'(x) = 3x^2 - 48(1) + 0$ (A1) for correct derivatives

$f'(x) = 3x^2 - 48$ A1   N2

[2]

(c) $f'(x) > 0$

$3x^2 - 48 > 0$ (A1) for correct inequality

By considering the graph of $y = 3x^2 - 48$,

$x < -4$ or $x > 4$.   A2   N3

[3]

Exercise 56

1. Let $f(x) = 2x^3 - 33x^2 + 108x - 125$.

   (a) Find the expression of $f'(x)$.

   [3]

   (b) Find the interval of x such that $f(x)$ is decreasing.

   [3]

Your Practice Set – Applications and Interpretation for IBDP Mathematics

2. Let $f(x) = -x^4 + 20x^3 - 142x^2 + 420x + 600$.

   (a) Find the expression of $f'(x)$.

   [3]

   (b) Find the intervals of x such that $f(x)$ is increasing.

   [3]

3. The table below shows the behaviour of $f(x)$ and $f'(x)$:

	$x < 3$	$x = 3$	$3 < x < 11$	$x = 11$	$x > 11$
$f(x)$		5		2	
$f'(x)$	Positive	Zero	Negative	Zero	Positive

   (a) (i) Write down the equation of the tangent at $x = 3$.

   (ii) Write down the equation of the tangent at $x = 11$.

   [2]

   (b) Write down the intervals when $f(x)$ is increasing.

   [2]

   (c) Write down the coordinates of the minimum point of $f(x)$.

   [2]

   (d) State which functional value is smaller: $f(5)$ or $f(9)$.

   [1]

4. The table below shows the behaviour of $f(x)$ and $f'(x)$:

	$x < -4$	$x = -4$	$-4 < x < 1$	$x = 1$	$1 < x < 5$	$x = 5$	$x > 5$
$f(x)$		0		20		12	
$f'(x)$	Negative	Zero	Positive	Zero	Negative	Zero	Negative

   (a) (i) Write down the equation of the tangent at $x = -4$.

   (ii) Write down the equation of the tangent at $x = 5$.

   [2]

   (b) Write down the intervals when $f(x)$ is decreasing.

   [2]

   (c) Write down the coordinates of the maximum point of $f(x)$.

   [2]

   (d) State which functional value is the largest: $f(2)$, $f(5)$ or $f(8)$.

   [1]

Paper 1 – Optimization Problems

Example

In a shop, the profit of selling x laptops can be modelled by $P(x) = 12x^3 - 288x^2 + 1728x$, where $0 \leq x \leq 10$.

(a) Find $P'(x)$.

[2]

(b) Hence, find the number of laptops to be sold such that the profit is maximized.

[3]

(c) Write down the exact maximum profit.

[1]

Solution

(a) $P'(x) = 12(3x^2) - 288(2x) + 1728(1)$ (A1) for correct derivative
$P'(x) = 36x^2 - 576x + 1728$ A1   N2

[2]

(b) $P'(x) = 0$
$36x^2 - 576x + 1728 = 0$ (M1) for setting equation
By considering the graph of
$y = 36x^2 - 576x + 1728$, $x = 4$ or
$x = 12$ (*Rejected*).   (M1) for valid approach
Thus, the required number of laptops is 4.   A1   N3

[3]

(c) $3072   A1   N1

[1]

Exercise 57

1. In a factory, the average cost of producing a titanium ring of x kilograms can be modelled by $C(x) = x^2 + 6 + \dfrac{54}{x}$, where $1 \leq x \leq 7$.

   (a) Find $C'(x)$.

   [2]

   (b) Hence, find the mass of a titanium ring to be produced such that the average cost is minimized.

   [3]

(c) Write down the value of the minimum average cost.

[1]

2. In an experiment about bacterial growth, the population of bacteria (in million) in a test tube can be modelled by $P(t) = t^3 - 12t^2 + 36t + 0.001$, where t is the number of days after the start of the experiment. The experiment ends after one week.

(a) Write down the initial amount of bacteria.

[1]

(b) Find $P'(t)$.

[2]

It is given that $P(7) = 7.001$.

(c) Find the number of days after the start of the experiment when the population of bacteria reaches its maximum.

[3]

(d) Write down the value of the maximum population of bacteria.

[1]

3. The temperature Q of a substance at time t is given by $Q(t) = 4t^2 - 120 + \dfrac{216}{t}$, where $t > 0$.

(a) Find the t-intercepts of Q.

[3]

(b) Find $Q'(t)$.

[2]

(c) Find the value of t when the temperature of the substance reaches its minimum.

[3]

4. The price P of a share at time t is given by $P(t) = -t^3 + 9t^2 - 24t + 720$, where $0 \leq t \leq 3$.

(a) Find $P'(t)$.

[2]

(b) Find the value of t when the price of the share reaches its minimum.

[3]

(c) State the reason why the price of the share cannot be lower than $690.

[1]

58 Paper 2 – Analysis of Graphs of Functions

Example

Let $f(x) = 2x^3 - 27x^2 + 108x + 27$. The graph of f has a local minimum at $(6, r)$.

(a) Find the value of r.

[2]

(b) Find $f'(x)$.

[2]

(c) Find the coordinates of the local maximum point of the graph of f.

[3]

(d) Write down the interval of x when $f(x)$ is decreasing.

[2]

(e) (i) Write down the value of $f(4)$.

    (ii) Find the value of $f'(4)$.

    (iii) Hence, find the equation of tangent at $x = 4$, giving the answer in general form.

[5]

Solution

(a) $r = 2(6)^3 - 27(6)^2 + 108(6) + 27$ (A1) for substitution
       $r = 135$ A1   N2

[2]

(b) $f'(x) = 2(3x^2) - 27(2x) + 108(1) + 0$ (A1) for correct derivatives
       $f'(x) = 6x^2 - 54x + 108$ A1   N2

[2]

(c) $f'(x) = 0$
       $6x^2 - 54x + 108 = 0$ (M1) for setting equation
       $6(x-3)(x-6) = 0$
       $x = 3$ or $x = 6$ (*Rejected*)   (A1) for correct value
       $f(3) = 2(3)^3 - 27(3)^2 + 108(3) + 27$
       $f(3) = 162$
       Thus, the required coordinates are $(3, 162)$.   A1   N3

[3]

Your Practice Set – Applications and Interpretation for IBDP Mathematics

    (d) $3 < x < 6$ A2   N2

                                                                                                                                                                   [2]

    (e)   (i)   155   A1   N1

            (ii) $f'(4) = 6(4)^2 - 54(4) + 108$ (M1) for substitution
                      $f'(4) = -12$ A1   N2

            (iii) The equation of tangent:
                      $y - 155 = -12(x - 4)$ (M1) for substitution
                      $y - 155 = -12x + 48$
                      $12x + y - 203 = 0$ A1   N2

                                                                                                                                          [5]

Exercise 58

1.   Let $f(x) = -x^3 + 3x^2 + 24x - 1$. The graph of f has a local minimum at $(-2, r)$.

    (a)   Find the value of r.

                                                                                                                                                              [2]

    (b)   Find $f'(x)$.

                                                                                                                                                              [2]

    (c)   Find the coordinates of the local maximum point of the graph of f.

                                                                                                                                                              [3]

    (d)   Write down the intervals of x when $f(x)$ is decreasing.

                                                                                                                                                              [2]

    (e)   (i)   Write down the value of $f(0)$.

            (ii)   Find the value of $f'(0)$.

            (iii) Hence, find the equation of tangent at $x = 0$, giving the answer in general form.

            (iv) Find the x-intercept of the tangent.

                                                                                                                                          [7]

2.   Let $f(x) = 2x^3 - 150x - 150$. The graph of f has a local maximum at $(r, 350)$.

    (a)   Find the value of r.

                                                                                                                                                              [2]

    (b)   Find $f'(x)$.

                                                                                                                                                              [2]

(c) Find the coordinates of the local minimum point of the graph of f.

[3]

(d) Write down the intervals of x when $f(x)$ is increasing.

[2]

(e) (i) Write down the value of $f(-1)$.

(ii) Find the value of $f'(-1)$.

(iii) Hence, write down the gradient of the line perpendicular to the tangent at $x = -1$.

(iv) Find the equation of normal at $x = -1$, giving the answer in slope-intercept form.

(v) Find the x-intercept of the normal.

[8]

3. Let $f(x) = 125x + \dfrac{32}{x^2}$, where $x \neq 0$.

(a) Write down the equation of the vertical asymptote of the graph of f.

[1]

(b) Find $f'(x)$.

[2]

(c) Find the coordinates of the local minimum point of the graph of f.

[3]

(d) Write down the interval of x when $f(x)$ is decreasing.

[2]

(e) (i) Write down the value of $f(1)$.

(ii) Find the value of $f'(1)$.

(iii) Hence, find the equation of tangent at $x = 1$, giving the answer in general form.

[5]

(f) The equation $f(x) = k$ has only one solution. Write down the range of values of k.

[2]

Your Practice Set – Applications and Interpretation for IBDP Mathematics

4. Let $f(x) = \dfrac{1}{2}x^2 + \dfrac{8}{x^2}$, where $x \neq 0$ and $x \geq -1$.

   (a) Write down the equation of the vertical asymptote of the graph of f.

   [1]

   (b) Find $f'(x)$.

   [2]

   (c) Find the coordinates of the local minimum point of the graph of f.

   [3]

   (d) Write down the intervals of x when $f(x)$ is increasing.

   [2]

   (e) (i) Write down the value of $f(4)$.

   (ii) Find the value of $f'(4)$.

   (iii) Hence, write down the gradient of the line perpendicular to the tangent at $x = 4$.

   (iv) Find the equation of normal at $x = 4$, giving the answer in slope-intercept form.

   [6]

   (f) The equation $f(x) = k$ has no solution. Write down the range of values of k.

   [2]

Paper 2 – Optimization Problems involving 3-D Objects

Example

A toy in the shape of circular cone is produced such that the sum of its height h and its radius r is 45 cm. Let V be the volume of the toy.

(a) Show that $V = 15\pi r^2 - \dfrac{1}{3}\pi r^3$.

[2]

(b) Find $\dfrac{dV}{dr}$.

[2]

(c) Find the radius of the toy when the volume reaches its maximum.

[3]

(d) Find the maximum volume of the toy in terms of π.

[2]

(e) Find the total surface area of the toy when its volume reaches its maximum.

[3]

(f) Find the values of the radii when the volume of the toy is 3000π cm^3.

[3]

Solution

(a) $h + r = 45$
$h = 45 - r$ M1
$V = \dfrac{1}{3}\pi r^2 h$
$V = \dfrac{1}{3}\pi r^2 (45 - r)$ A1
$V = 15\pi r^2 - \dfrac{1}{3}\pi r^3$ AG   N0

[2]

(b) $\dfrac{dV}{dr} = 15\pi(2r) - \dfrac{1}{3}\pi(3r^2)$ (A1) for correct derivatives

$\dfrac{dV}{dr} = 30\pi r - \pi r^2$ A1   N2

[2]

Your Practice Set – Applications and Interpretation for IBDP Mathematics

(c) $\dfrac{dV}{dr} = 0$

$30\pi r - \pi r^2 = 0$ (M1) for setting equation

By considering the graph of $y = 30\pi r - \pi r^2$,

$r = 0$ (Rejected) or $r = 30$.   (M1) for valid approach

Thus, the required radius is 30 cm.   A1   N3

[3]

(d)   The maximum volume

$= 15\pi(30)^2 - \dfrac{1}{3}\pi(30)^3$ (A1) for substitution

$= 4500\pi$ cm^3 A1   N2

[2]

(e)   The total surface area

$= \pi(30)^2 + \pi(30)(\sqrt{30^2 + 15^2})$ (M1)(A1) for correct approach

$= 5988.600017$

$= 5990$ cm^2 A1   N3

[3]

(f) $V = 3000\pi$

$15\pi r^2 - \dfrac{1}{3}\pi r^3 = 3000\pi$ (M1) for setting equation

$15\pi r^2 - \dfrac{1}{3}\pi r^3 - 3000\pi = 0$

By considering the graph of

$y = 15\pi r^2 - \dfrac{1}{3}\pi r^3 - 3000\pi$, $r = 18.391106$ or

$r = 39.118693$.   (M1) for valid approach

Thus, the required radii are 18.4 cm and 39.1 cm.   A1   N3

[3]

Exercise 59

1.   A cylindrical blueberry can of height h and radius r, with an open end removed, is produced such that the total outer surface area A is 27π cm^2. Let V be the volume of the can.

     (a)   (i)   Express h in terms of r.

            (ii)   Show that $V = \dfrac{27}{2}\pi r - \dfrac{1}{2}\pi r^3$.

[4]

(b) Find $\dfrac{dV}{dr}$.

[2]

(c) Find the radius of the can when the volume reaches its maximum.

[3]

(d) (i) Find the maximum volume of the can in terms of π.

(ii) Hence, write down the range of V.

[3]

(e) Find the values of the radii when the volume of the can is $47\,\text{cm}^3$.

[3]

2. In a Design & Technology lesson, a teacher uses a square paper with length 64 cm to make an open box. Firstly, four squares with lengths x cm on four corners are removed. Then the four rectangular flaps are folded up to form an open box, such that the base is a square.

(a) (i) Write down the length of the square base of the open box.

(ii) Show that $V = 4x^3 - 256x^2 + 4096x$.

[3]

(b) Find $\dfrac{dV}{dx}$.

[2]

(c) Find the value of x when the volume reaches its maximum.

[3]

(d) (i) Find the maximum volume of the open box, giving the answer correct to the nearest cm^3.

(ii) Hence, write down the range of V.

[3]

(e) All surfaces of the open box has to be painted afterwards. Find the total surface area that has to be painted when its volume reaches its maximum.

[3]

3. A box in the shape of triangular prism with height h is produced such that the total surface area A is $100\,\text{cm}^2$. The cross-sectional area of the box is an equilateral triangle with side length r. Let V be the volume of the box.

(a) (i) Using $\sin 60° = \dfrac{\sqrt{3}}{2}$, show that $h = \dfrac{200 - \sqrt{3}r^2}{6r}$.

Your Practice Set – Applications and Interpretation for IBDP Mathematics

      (ii)   Show that $V = \dfrac{25\sqrt{3}}{3}r - \dfrac{1}{8}r^3$.

[5]

(b)   Find $\dfrac{dV}{dr}$.

[2]

(c)   Find the value of r when the volume reaches its maximum.

[3]

(d)   Find the maximum volume of the box.

[2]

The density of a three-dimensional solid is defined as the ratio of its mass to its volume. It is given that the mass of the box when its volume reaches its maximum, is 9 kg.

(e)   Find the density of this box, giving the answer in the form $a \times 10^k$, where $1 \leq a < 10$ and $k \in \mathbb{Z}$.

[2]

4.   An acrylic block in the shape of prism with height h is produced such that the volume V is 168 cm^3. The cross-sectional area of the block is a quarter circle with radius r. Let A be the total surface area of the block.

(a)   (i)   Express h in terms of r.

        (ii)   Hence, show that $A = \dfrac{1}{2}\pi r^2 + \left(\dfrac{1344}{\pi} + 336\right)\dfrac{1}{r}$.

[5]

(b)   Find $\dfrac{dA}{dr}$, giving the answer in terms of π.

[2]

(c)   Find the value of r when the total surface area reaches its minimum.

[3]

(d)   Find the minimum total surface area of the block.

[2]

For an acrylic block with the total surface area reaches its minimum, painting buckets are used to make paint on the surface of the block. It is given that each bucket contains an amount of paint which can cover 25 cm^2 of the surface.

(e)   Find the minimum number of painting buckets required.

[2]

Chapter 16

Integration and Trapezoidal Rule

SUMMARY POINTs

- Integrals of a function $y = f(x)$:
    1. $\int f(x)\,dx$: Indefinite integral of $f(x)$
    2. $\int_a^b f(x)\,dx$: Definite integral of $f(x)$ from a to b

- Rules of integration:
    1. $\int x^n\,dx = \dfrac{1}{n+1} x^{n+1} + C$
    2. $\int (p'(x) + q'(x))\,dx = p(x) + q(x) + C$
    3. $\int cp'(x)\,dx = cp(x) + C$

- $\int_a^b f(x)\,dx$: Area under the graph of $f(x)$ and above the x-axis, between $x = a$ and $x = b$

Your Practice Set – Applications and Interpretation for IBDP Mathematics

SUMMARY POINTs

✓ Trapezoidal Rule:

$a, b\ (a < b)$: End points

n: Number of intervals

$h = \dfrac{b-a}{n}$: Interval width

$\int_a^b f(x)\,dx$ can be estimated by $\dfrac{1}{2}h[f(x_0) + f(x_n) + 2(f(x_1) + f(x_2) + \cdots + f(x_{n-1}))]$

✓ Estimation by Trapezoidal Rule:
1. The estimation overestimates if the estimated value is greater than the actual value of $\int_a^b f(x)\,dx$
2. The estimation underestimates if the estimated value is less than the actual value of $\int_a^b f(x)\,dx$

 Solutions of Chapter 16

Paper 1 – Finding the Original Functions

Example

The derivative of f is given by $f'(x) = 3x^2 + 3x - 18$. The y-intercept of the graph is at $(0, -3)$. Find an expression for $f(x)$.

[5]

Solution

$f(x) = \int (3x^2 + 3x - 18)\,dx$ (M1) for indefinite integral

$f(x) = 3\left(\dfrac{x^3}{3}\right) + 3\left(\dfrac{x^2}{2}\right) - 18x + C$ (A2) for correct integration

$f(x) = x^3 + \dfrac{3}{2}x^2 - 18x + C$

$-3 = 0^3 + \dfrac{3}{2}(0)^2 - 18(0) + C$ (M1) for substitution

$C = -3$

$\therefore f(x) = x^3 + \dfrac{3}{2}x^2 - 18x - 3$ A1 N5

[5]

Exercise 60

1. The derivative of f is given by $f'(x) = 6x^2 - 2x - 8$. The y-intercept of the graph is at $(0, 7)$. Find an expression for $f(x)$.

    [5]

2. The derivative of f is given by $f'(x) = x^2 - 36 - \dfrac{3}{x^2}$, where $x \neq 0$. The x-intercept of the graph is at $(3, 0)$.

    (a) Find an expression for $f(x)$.

    [5]

    (b) The graph of f passes through $\left(a, \dfrac{377}{8}\right)$, where $0 < a < 4$. Find the value of a.

    [2]

Your Practice Set – Applications and Interpretation for IBDP Mathematics

3. The derivative of f is given by $f'(x) = x^2 - 64x + 12$. The x-intercept of the graph is at A(6, 0).

    (a) Find an expression for $f(x)$.
    [5]

    (b) The graph of f passes through B(0, b). Write down the value of b.
    [1]

    (c) Hence, find the area of the triangle OAB, where O is the origin.
    [2]

4. The derivative of f is given by $f'(x) = -\dfrac{200}{x^3} - \dfrac{20}{x^2}$, where $x \neq 0$. The graph of f passes through the point A(−10, 0).

    (a) Find an expression for $f(x)$.
    [5]

    The point B is a point above the x-axis with y-coordinate b. The area of the triangle OAB is 100, where O is the origin.

    (b) Find the value of b.
    [3]

61 Paper 1 – Area Under a Curve

Example

Consider the graph of $f(x) = 2(x-2)(8-x)$, where $f(x) \geq 0$ for $a \leq x \leq b$. Let R be the region enclosed by the graph of f and the x-axis.

(a) (i) Write down the values of a and b.

(ii) Write down the equation of the axis of symmetry of $f(x)$.

[3]

(b) (i) Write down the integral representing the area of R.

(ii) Hence, find the area of R.

[2]

Solution

(a) (i) $a = 2$, $b = 8$ A2   N2

(ii) $x = 5$ A1   N1

[3]

(b) (i) $\int_2^8 2(x-2)(8-x)\,dx$ A1   N1

(ii) 72   A1   N1

[2]

Exercise 61

1. Consider the graph of $f(x) = -x^3 + 16x^2 - 69x + 90$ for $x \geq 3$, where $f(x) \geq 0$ for $3 \leq x \leq p$. Let R be the region enclosed by the graph of f and the x-axis.

   (a) (i) Write down the value of p.

   (ii) Write down the coordinates of the maximum point of $f(x)$.

   [3]

   (b) (i) Write down the integral representing the area of R.

   (ii) Hence, find the exact value of the area of R.

   [2]

Your Practice Set – Applications and Interpretation for IBDP Mathematics

2. Consider the graph of $f(x) = x^3 - 5x^2 - 25x + 125$ for $x \leq 6$, where $f(x) \geq 0$ for $p \leq x \leq 6$. Let R be the region enclosed by the graph of f, the y-axis and the x-axis.

   (a) (i) Write down the value of p.

   (ii) Write down the coordinates of the minimum point of $f(x)$.

   [3]

   (b) (i) Write down the integral representing the area of R.

   (ii) Hence, find the exact value of the area of R.

   [2]

3. Consider the graph of $f(x) = -2x^2 - 8x + 42$, where $f(x) \geq 0$ for $a \leq x \leq b$. Let R be the region enclosed by the graph of f, the line $x = c$ and the x-axis.

   (a) (i) Write down the values of a and b.

   (ii) Write down the coordinates of the maximum point of $f(x)$.

   [4]

   The area of R can be calculated by $\int_a^c (-2x^2 - 8x + 42) dx$, where $a \leq c \leq b$. It is given that the area of R is $\dfrac{28}{3}$.

   (b) Find the value of c.

   [2]

4. Consider the graph of $f(x) = -2x^2 + 120x - 1600$, where $f(x) \geq 0$ for $a \leq x \leq b$. Let R be the region enclosed by the graph of f and the x-axis.

   (a) (i) Write down the values of a and b.

   (ii) Write down the integral representing the area of R.

   (iii) Hence, find the exact value of the area of R.

   [4]

   Consider the parallelogram PQRS with vertices $P(17, c)$, $Q(117, c)$, $R(92, 30)$ and $S(-8, 30)$, where $c > 30$. It is given that the area of the parallelogram is equal to the area of R.

   (b) Find the value of c.

   [3]

62 Paper 1 – Integration and 3-D Objects

Example

Consider the graph of $f(x) = 3(x-1)(5-x)$, where $f(x) \geq 0$ for $1 \leq x \leq 5$. Let R be the region enclosed by the graph of f and the x-axis.

(a) (i) Write down the integral representing the area of R.

    (ii) Hence, find the area of R.

[2]

A triangular prism is formed with height 7.5 and the cross-sectional area equals to the area of R.

(b) Find the volume of the triangular prism.

[2]

Solution

(a) (i) $\int_1^5 3(x-1)(5-x)\,dx$ A1   N1

    (ii) 32   A1   N1

[2]

(b) The volume of the triangular prism
$= (32)(7.5)$ (A1) for substitution
$= 240$ A1   N2

[2]

Exercise 62

1. Consider the graph of $f(x) = 1.5(x+2)(x-10)^2$, where $f(x) \geq 0$ for $-2 \leq x \leq 10$. Let R be the region enclosed by the graph of f and the x-axis.

   (a) (i) Write down the integral representing the area of R.

       (ii) Hence, find the exact area of R.

   [2]

Your Practice Set – Applications and Interpretation for IBDP Mathematics

A square-based pyramid is formed with height 15 and the base area equals to the area of R.

(b)   Find the exact volume of the pyramid.

[2]

2. Consider the graph of $f(x) = \pi(9-(x-3)^2)$, where $f(x) \geq 0$ for $0 \leq x \leq 6$. Let R be the region enclosed by the graph of f and the x-axis.

   (a)   Find the area of R in terms of π.

[2]

   A sphere is formed such that its total surface area equals to the area of R.

   (b)   Find the radius of the sphere.

[2]

3. Consider the graph of $f(x) = (x-4)^2(8-x)$, where $f(x) \geq 0$ for $4 \leq x \leq 8$. Let R be the region enclosed by the graph of f and the x-axis.

   (a)   (i)   Write down the integral representing the area of R.

          (ii)   Hence, find the exact area of R.

[2]

   A prism is formed with height h and the cross-sectional area equals to the area of R. The volume of the prism is 320.

   (b)   Find the value of h.

[2]

4. Consider the graph of $f(x) = (x-8)^2(x-16)^2$, where $f(x) \geq 0$ for $12 \leq x \leq 16$. Let R be the region enclosed by the graph of f and the x-axis.

   (a)   Find the exact area of R.

[2]

   A pyramid is formed with height h and the base area equals to five times the area of R. The volume of the pyramid is 1024.

   (b)   Find the exact value of h.

[2]

Paper 1 – Trapezoidal Rule

Example

Consider the integral $\int_0^{12} (\sqrt[3]{x})dx$. The value of this integral is estimated by trapezoidal rule with 6 trapeziums.

(a) Find the width of each trapezium.

[2]

(b) Hence, find an estimate of $\int_0^{12} (\sqrt[3]{x})dx$.

[3]

Solution

(a) The width of each trapezium
$= \dfrac{12-0}{6}$ (A1) for correct substitution
$= 2$ A1   N2

[2]

(b) The estimate of $\int_0^{12} (\sqrt[3]{x})dx$

$= \dfrac{1}{2}(2)\left[\sqrt[3]{0} + \sqrt[3]{12} + 2(\sqrt[3]{2} + \sqrt[3]{4} + \sqrt[3]{6} + \sqrt[3]{8} + \sqrt[3]{10})\right]$ (A2) for substitution

$= 19.92718325$

$= 19.9$ A1   N3

[3]

Exercise 63

1. Consider the integral $\int_1^7 x^{0.3} dx$. The value of this integral is estimated by trapezoidal rule with 4 trapeziums.

    (a) Find the width of each trapezium.

    [2]

    (b) Hence, find an estimate of $\int_1^7 x^{0.3} dx$.

    [3]

Your Practice Set – Applications and Interpretation for IBDP Mathematics

2. Consider the integral $\int_{0.5}^{0.8} 4^x \, dx$. The value of this integral is estimated by trapezoidal rule with n trapeziums, such that the width of each trapezium is 0.1.

   (a) Find the value of n.

   [2]

   (b) Find an estimate of $\int_{0.5}^{0.8} 4^x \, dx$.

   [3]

3. Consider the integral $\int_{a}^{6} e^x \, dx$. The value of this integral is estimated by trapezoidal rule with 6 trapeziums, such that the width of each trapezium is 0.4.

   (a) Find the value of a.

   [2]

   (b) Hence, find an estimate of $\int_{a}^{6} e^x \, dx$.

   [3]

4. Consider the function $f(x) = \dfrac{1}{2x}$ for $x > 0$. The area of the region enclosed by the graph of f, the line $x = b$, the x-axis and the line $x = 2$, where $b > 2$, is estimated by trapezoidal rule with 8 trapeziums, such that the width of each trapezium is 0.25.

   (a) Write down the integral representing the area of the region.

   [1]

   (b) Find the value of b.

   [2]

   (c) Hence, find an estimate of area of the region.

   [3]

Paper 1 – Errors in Trapezoidal Rule

Example

Consider the integral $\int_0^6 \frac{2}{(1+x)^2}\,dx$. The value of this integral is estimated by trapezoidal rule with 6 trapeziums, such that the width of each trapezium is 1.

(a) Find an estimate of $\int_0^6 \frac{2}{(1+x)^2}\,dx$.

[3]

It is given that the exact value of $\int_0^6 \frac{2}{(1+x)^2}\,dx$ is $\frac{12}{7}$.

(b) Calculate the percentage error in the estimate.

[2]

Solution

(a) The estimate of $\int_0^6 \frac{2}{(1+x)^2}\,dx$

$$= \frac{1}{2}(1)\left[\frac{2}{(1+0)^2} + \frac{2}{(1+6)^2} + 2\left(\frac{2}{(1+1)^2} + \frac{2}{(1+2)^2} + \frac{2}{(1+3)^2} + \frac{2}{(1+4)^2} + \frac{2}{(1+5)^2}\right)\right]$$

(A2) for substitution

$= 2.003185941$

$= 2.00$ A1   N3

[3]

(b) The percentage error

$$= \left|\frac{2.003185941 - \frac{12}{7}}{\frac{12}{7}}\right| \times 100\%$$

(A1) for correct substitution

$= 16.85251323\%$

$= 16.9\%$ A1   N2

[2]

Your Practice Set – Applications and Interpretation for IBDP Mathematics

Exercise 64

1. Consider the integral $\int_{-11}^{-8} \frac{3}{\sqrt{x+12}} dx$. The value of this integral is estimated by trapezoidal rule with 4 trapeziums, such that the width of each trapezium is 0.75.

    (a) Find an estimate of $\int_{-11}^{-8} \frac{3}{\sqrt{x+12}} dx$.

    [3]

    It is given that the exact value of $\int_{-11}^{-8} \frac{3}{\sqrt{x+12}} dx$ is 6.

    (b) Calculate the percentage error in the estimate.

    [2]

2. Consider the integral $\int_{0}^{3} 4e^x dx$. The value of this integral is estimated by trapezoidal rule with 5 trapeziums.

    (a) Write down the width of each trapezium.

    [1]

    (b) Hence, find an estimate of $\int_{0}^{3} 4e^x dx$.

    [3]

    It is given that the exact value of $\int_{0}^{3} 4e^x dx$ is $4(e^3 - 1)$.

    (c) Calculate the percentage error in the estimate.

    [2]

3. The following table shows the functional values of $f(x)$ from $x = 5$ to $x = 6$:

x	5	5.125	5.25	5.375	5.5	5.625	5.75	5.875	6
$f(x)$	8	7	6	5	4	3	$f(5.75)$	$f(5.875)$	8

    It is given that $f(5.625)$, $f(5.75)$, $f(5.875)$ and $f(6)$ forms an arithmetic sequence.

    (a) (i) Write down the common difference of the an arithmetic sequence.

    (ii) Hence, write down the exact values of $f(5.75)$ and $f(5.875)$.

    [3]

    (b) Find an estimate of $\int_{5}^{6} f(x) dx$.

    [3]

It is given that the exact value of $\int_5^6 f(x)\,dx$ is 5.2.

(c) State whether the estimate in (b) overestimates or underestimates $\int_5^6 f(x)\,dx$.

[1]

4. The following table shows the functional values of $f(x) = 0.25^x$ from $x = 0.3$ to $x = 0.6$:

x	0.3	a	0.4	0.45	0.5	0.55	0.6
$f(x)$	$0.25^{0.3}$	0.25^a	$0.25^{0.4}$	$0.25^{0.45}$	b	$0.25^{0.55}$	$0.25^{0.6}$

(a) Write down the exact values of a and b.

[2]

(b) Find an estimate of $\int_{0.3}^{0.6} f(x)\,dx$.

[3]

It is given that the exact value of $\int_{0.3}^{0.6} f(x)\,dx$ is 0.1619271347.

(c) State whether the estimate in (b) overestimates or underestimates $\int_{0.3}^{0.6} f(x)\,dx$.

[1]

Your Practice Set – Applications and Interpretation for IBDP Mathematics

Chapter

Statistics

SUMMARY POINTs

✓ Relationship between frequencies and cumulative frequencies:

Data	Frequency	Data less than or equal to	Cumulative frequency
10	f_1	10	f_1
20	f_2	20	$f_1 + f_2$
30	f_3	30	$f_1 + f_2 + f_3$

✓ Measures of central tendency for a data set $\{x_1, x_2, x_3, \cdots, x_n\}$ arranged in ascending order:

1. $\bar{x} = \dfrac{x_1 + x_2 + x_3 + \ldots + x_n}{n}$: Mean
2. The datum or the average value of two data at the middle: Median
3. The datum appear the most: Mode

SUMMARY POINTs

✓ Measures of dispersion for a data set $\{x_1, x_2, x_3, \cdots, x_n\}$ arranged in ascending order:
   1. $x_n - x_1$: Range
   2. Two subgroups A and B can be formed from the data set such that all data of the subgroup A are less than or equal to the median, while all data of the subgroup B are greater than or equal to the median
   3. Q_1 = The median of the subgroup A: Lower quartile
   4. Q_3 = The median of the subgroup B: Upper quartile
   5. $Q_3 - Q_1$: Inter-quartile range (IQR)
   6. $\sigma = \sqrt{\dfrac{(x_1 - \overline{x})^2 + (x_2 - \overline{x})^2 + (x_3 - \overline{x})^2 + \ldots + (x_n - \overline{x})^2}{n}}$: Standard deviation

✓ Box-and-whisker diagram:

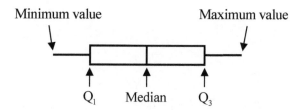

✓ A datum x is defined to be an outlier if $x < Q_1 - 1.5\text{IQR}$ or $x > Q_3 + 1.5\text{IQR}$

Coding of data:
   1. Only the mean, the median, the mode and the quartiles will change when each datum of the data set is added or subtracted by a value
   2. All measures of central tendency and measures of dispersion will change when each datum of the data set is multiplied or divided by a value

 Solutions of Chapter 17

Paper 1 – Measures of Central Tendency and Dispersion

Example

In an examination, the grades of a group of 11 students are as follows:

    7   6   4   5   7   4   7   6   2   1   6

(a) Find the mean of the grades. [2]

(b) Write down the median of the grades. [1]

(c) Write down the mode(s) of the grades. [2]

(d) Write down the standard deviation of the grades. [1]

Solution

(a) The mean
$$= \frac{7+6+4+5+7+4+7+6+2+1+6}{11}$$
$$= 5$$

(A1) for correct formula

A1   N2 [2]

(b) 6   A1   N1 [1]

(c) 6, 7   A2   N2 [2]

(d) 1.95   A1   N1 [1]

Exercise 65

1. The number of snowy days in a town in 2019 are as follows:

   8 9 14 18 2 5 1 0 0 10 6 11

   (a) Find the mean of the data.
   [2]

   (b) Write down the median of the data.
   [1]

   (c) Find the inter-quartile range of the data.
   [2]

   (d) Write down the standard deviation of the data.
   [1]

2. The number of nouns in each sentence of a newspaper passage are as follows:

   3 8 6 10 2 7 7 9

   (a) Find the mean of the data.
   [2]

   (b) Write down the value of

   (i) median;

   (ii) upper quartile;

   (iii) lower quartile;

   (iv) inter-quartile range.
   [4]

3. Consider the following data set of ten numbers:

   7 6 12 3 6 8 11 5 4 x

   The mean of the data set is 7.

   (a) Find the value of x.
   [2]

   (b) Write down the mode(s) of the data set.
   [2]

   (c) Find the inter-quartile range of the data set.
   [2]

Your Practice Set – Applications and Interpretation for IBDP Mathematics

4. The weights of 13 students (in kg) are shown in the following stem-and-leaf diagram:

Stem	Leaf					
4	5	6	7	9		
5	0	0	3	3	3	x
6	2	4	4			

    It is given that the inter-quartile range of the weights is 11.

    (a) (i) Write down the lower quartile of the weights.

    (ii) Write down the upper quartile of the weights.

    (iii) Hence, find the value of x.
    [4]

    (b) Write down the range of the weights.
    [1]

    (c) Write down the standard deviation of the data.
    [1]

Paper 1 – Frequency Tables of Continuous Data

Example

The table below shows the frequency distribution of the heights of 60 students:

Height (cm)	Frequency
$140 \leq x < 150$	5
$150 \leq x < 160$	10
$160 \leq x < 170$	17
$170 \leq x < 180$	25
$180 \leq x < 190$	3

(a)   (i)   Write down the class width for the class $140 \leq x < 150$.

      (ii)   Write down the class mark for the class $160 \leq x < 170$.

[2]

(b)   State whether the data is discrete or continuous.

[1]

(c)   Find an estimate for

      (i)   the mean height;

      (ii)   the standard deviation;

      (iii)   the variance.

[4]

Solution

(a)   (i)   10 cm   A1   N1

      (ii)   165 cm   A1   N1

[2]

(b)   Continuous   A1   N1

[1]

(c)   (i)   167 cm   A2   N2

      (ii)   10.4 cm   A1   N1

      (iii)   108 cm   A1   N1

[4]

Your Practice Set – Applications and Interpretation for IBDP Mathematics

Exercise 66

1. The table below shows the frequency distribution of the number of full stops in a document of 80 pages:

Number of fullstops	Frequency
$25 \leq x < 30$	18
$30 \leq x < 35$	34
$35 \leq x < 40$	11
$40 \leq x < 45$	10
$45 \leq x < 50$	7

   (a) (i) Write down the class mark for the class $25 \leq x < 30$.

       (ii) Write down the modal class.

       [2]

   (b) Find an estimate for

       (i) the mean number of full stops, giving the answer in exact value;

       (ii) the standard deviation;

       (iii) the variance.

       [4]

2. The table below shows the frequency distribution of the amount of money spent by 200 customers in an outlet store on a particular Sunday:

Amount of money (USD)	Frequency
$0 \leq x < 50$	87
$50 \leq x < 100$	55
$100 \leq x < 150$	27
$150 \leq x < 200$	11
$200 \leq x < 250$	f
$250 \leq x < 300$	5

   (a) Find the value of f.

       [2]

(b) (i) State whether the data is discrete or continuous.

(ii) Write down the class mark for the class $150 \leq x < 200$.

(iii) Write down the modal class.

[3]

(c) Find an estimate for

(i) the mean amount of money, giving the answer in exact value;

(ii) the standard deviation.

[3]

3. The table below shows the frequency distribution of the scores obtained from their internal assessment in Mathematics:

Scores	Frequency
$0 \leq x < 4$	2
$4 \leq x < 8$	3
$8 \leq x < f$	4
$f \leq x < 16$	6
$16 \leq x < 20$	3

It is given that the class widths of all classes are the same, and all scores are positive integers.

(a) Write down the value of f.

[1]

(b) (i) Write down the class mark for the class $16 \leq x < 20$.

(ii) Write down the modal class.

[2]

(c) Find an estimate for

(i) the mean score;

(ii) the standard deviation.

[3]

(d) Find the greatest possible score that is one standard deviation above the mean score.

[2]

Your Practice Set – Applications and Interpretation for IBDP Mathematics

4. Consider the following data set of sixteen data:

$$1\ 3\ 5\ 5\ 5\ 7\ 9\ 2\ 2\ 2\ 2\ 4\ 6\ 8\ 10\ 0$$

The data set can be presented by the following frequency table:

Data	Frequency
$0 \leq x < 3$	p
$3 \leq x < 6$	q
$6 \leq x < 9$	r
$9 \leq x < 12$	s

(a) Write down the values of p, q, r and s.

[2]

(b) Write down the modal class.

[1]

(c) (i) Using the data set, find the standard deviation.

    (ii) Using the frequency table, find an estimate for the standard deviation.

    (iii) Hence, find the percentage error in the estimate in (c)(ii).

[5]

Paper 1 – Cumulative Frequency Table

Example

The following table gives the examination scores for 60 students.

Score	Number of students	Cumulative frequency
20	15	15
40	30	45
60	5	p
80	q	60

(a) Find the value of

  (i) p;

  (ii) q.

[3]

(b) Find the mean score.

[2]

(c) Write down the standard deviation.

[1]

(d) State whether the data is discrete or continuous.

[1]

Solution

(a) (i) $p = 45 + 5$
    $p = 50$ A1   N1

  (ii) $q = 60 - 50$ (M1) for valid approach
    $q = 10$ A1   N2

[3]

(b) The mean score
$$= \frac{(20)(15)+(40)(30)+(60)(5)+(80)(10)}{60}$$ (A1) for correct formula
$= 43.33333333$
$= 43.3$ A1   N2

[2]

(c) 19.7   A1   N1
                                                      [1]

(d) Discrete   A1   N1
                                                      [1]

Exercise 67

1. The following table gives the number of notebooks kept by 50 students.

Number of notebooks	Number of students	Cumulative frequency
1	14	14
2	7	p
3	q	39
4	10	49
5	1	50

   (a) Find the value of

   (i) p;

   (ii) q.
                                                      [3]
   (b) Find the mean number of notebooks.
                                                      [2]
   (c) Write down the standard deviation.
                                                      [1]

2. In a school with 180 boys, each student is tested to see how many sit-up exercises (sit-ups) he can do in one minute. The results are given in the table below.

Number of sit-ups	Number of students	Cumulative frequency
22	32	32
23	21	53
24	37	p
25	25	115
26	q	165
27	15	180

(a) Find the value of

   (i) p ;

   (ii) q .
[3]

(b) Find the mean number of sit-ups.
[2]

(c) Write down the variance.
[1]

3. The following frequency distribution of marks has mean 4.24.

Marks	Number of students	Cumulative frequency
1	2	2
2	4	6
3	6	12
4	16	28
5	p	$p+28$
6	10	q

(a) Find the value of p .
[4]

(b) Find the value of q .
[2]

(c) State whether the data is discrete or continuous.
[1]

4. The following frequency distribution of marks has mean 17.8.

Marks	Number of students	Cumulative frequency
7	5	5
12	3	8
17	6	14
22	5	19
27	p	$p+19$

(a) Find the value of p .
[4]

(b) Find the upper quartile of the distribution.
[2]

Your Practice Set – Applications and Interpretation for IBDP Mathematics

 Paper 1 – Box-and-whisker plots

Example

The following box-and-whisker plot shows the number of emails sent by teachers in a college on a particular week.

0   42 50 53 59

(a)   Find the inter-quartile range.

[2]

(b)   One teacher sent p emails, where $p < 42$. Given that p is an outlier, find the greatest value of p.

[4]

Solution

(a)   The inter-quartile range
      $= 53 - 42$ (A1) for correct formula
      $= 11$ A1   N2

[2]

(b) $p < 42 - 1.5(11)$ (M1)(A1) for correct inequality
      $p < 25.5$ (A1) for correct value
      Thus, the greatest value of p is 25.   A1   N4

[4]

Exercise 68

1.   The following box-and-whisker plot shows the number of parcels received by a group of youngsters on a particular week.

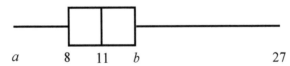

     a 8   11 b 27

The range and the inter-quartile range of the distribution are 24 and 6 respectively.

(a)   Write down the values of a and b.

[2]

(b) One youngster received p parcels, where $p > b$. Given that p is an outlier, find the least value of p.

[4]

2. The following box-and-whisker plot shows the test scores of a group of students in a particular test.

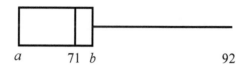

The range and the inter-quartile range of the distribution are 29 and 10 respectively.

(a) Write down the values of a and b.

[2]

(b) One student got k marks, where $k > b$. Given that k is an outlier, find the least value of k.

[4]

3. Consider the following box-and-whisker plot of a distribution.

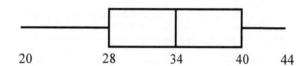

(a) Write down the values of

(i) median;

(ii) range;

(iii) inter-quartile range.

[3]

Consider the following table of the same distribution.

Data	Frequency
20	2
28	4
40	q
44	1

(b) Find the value of q.

[3]

4. Consider the following box-and-whisker plot of a distribution.

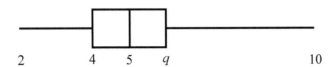

The inter-quartile range of the distribution is 2.

(a) Write down the values of

(i) median;

(ii) range;

(iii) q.

[3]

Consider the following table of the same distribution.

Data	Frequency
2	1
4	r
6	5
8	3
10	2

(b) Find the value of r.

[3]

69 Paper 1 – Effects of Coding in Data

Example

There are 20 items in a data set. The sum of the items is 100.

(a) Find the mean.

[2]

The variance of this data set is 2. Each value in the set is multiplied by 7.

(b) (i) Write down the value of the new mean.

(ii) Find the value of the new variance.

[3]

Solution

(a) The mean
$= \dfrac{100}{20}$ (A1) for correct formula
$= 5$ A1   N2

[2]

(b) (i) 35   A1   N1

(ii) The new variance
$= (7^2)(2)$ (M1) for valid approach
$= 98$ A1   N2

[3]

Exercise 69

1. There are 15 items in a data set. The sum of the items is 150.

    (a) Find the mean.

    [2]

    The variance of this data set is 8. Each value in the set is multiplied by 3.

    (b) (i) Write down the value of the new mean.

    (ii) Find the value of the new variance.

    [3]

Your Practice Set – Applications and Interpretation for IBDP Mathematics

2. There are 12 items in a data set. The mean of the items is 9.

   (a) Find the sum of the items.
   [2]

   The variance of this data set is 2.25. Each value in the set is added by 10.

   (b) (i) Write down the value of the new mean.

   (ii) Find the value of the new standard deviation.
   [3]

3. There are 16 items in a data set and arranged in ascending order. The 12th item and the 13th item are 20 and 22 respectively.

   (a) Find the upper quartile.
   [2]

   The lower quartile of this data set is 10. Each value in the set is multiplied by 4.

   (b) (i) Write down the value of the new lower quartile.

   (ii) Find the value of the new inter-quartile range.
   [3]

4. There are 25 items in a data set and arranged in ascending order. The 6th item and the 7th item are 8 and 12 respectively.

   (a) Find the lower quartile.
   [2]

   The inter-quartile range of this data set is 19. Each value in the set is added by 5.

   (b) (i) Write down the value of the new inter-quartile range.

   (ii) Find the value of the new upper quartile.
   [3]

Paper 2 – Cumulative Frequency Curves

Example

A city hired 80 workers to work at an event. The following cumulative frequency curve shows the number of hours workers worked during the event.

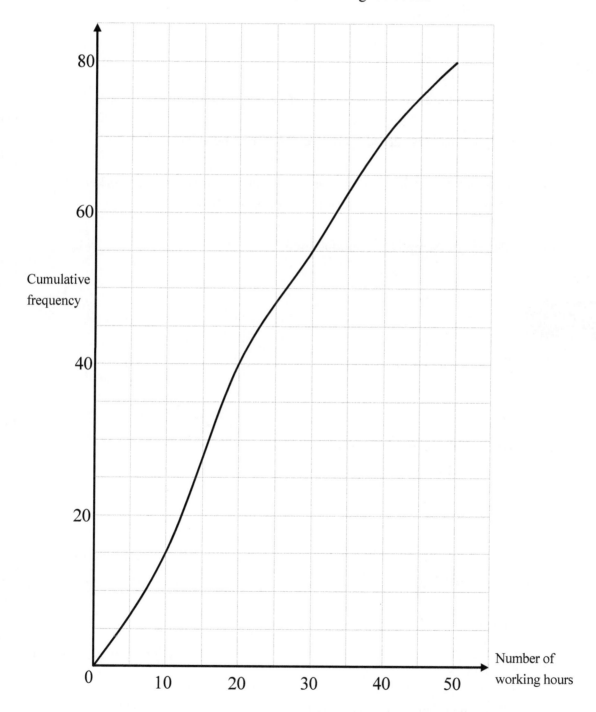

Your Practice Set – Applications and Interpretation for IBDP Mathematics

    (a)   (i)   Write down the median number of hours worked by workers.

            (ii)   Write down the number of workers who worked 30 hours or less.

[3]

The city paid each of the workers $20 per hour for the first 30 hours worked, and $30 per hour for each hour they worked after the first 30 hours.

    (b)   Find the amount of money a worker earned for working

            (i)   25 hours;

            (ii)   35 hours.

[4]

    (c)   Find the number of workers who earned less than or equal to $200.

[3]

    (d)   Only 10 workers earned more than k. Find the value of k.

[4]

The data above was obtained from a survey. A manager arranged the worker IDs in ascending order and then asked every 10th worker on the list.

    (e)   Identify the sampling technique used.

[1]

Solution

(a) (i) 20 hours   A2   N2

      (ii) 55   A1   N1

[3]

(b) (i) The amount of money
$= (20)(25)$
$= \$500$ A1   N1

      (ii) The amount of money
$= (20)(30) + (30)(35 - 30)$ (M1)(A1) for correct formula
$= \$750$ A1   N3

[4]

(c) The number of working hours
$= \dfrac{200}{20}$ (A1) for correct formula
$= 10$
Thus, the number of workers is 15.   A2   N3

[3]

(d) The number of workers earned not more than k
$= 80 - 10$ (M1) for valid approach
$= 70$ (A1) for correct value
$k = (20)(30) + (30)(40 - 30)$ (A1) for correct formula
$k = 900$ A1 N4

[4]

(e) Systematic sampling A1 N1

[1]

Exercise 70

1. The following cumulative frequency curve shows the amounts in dollars raised by all the students in the school.

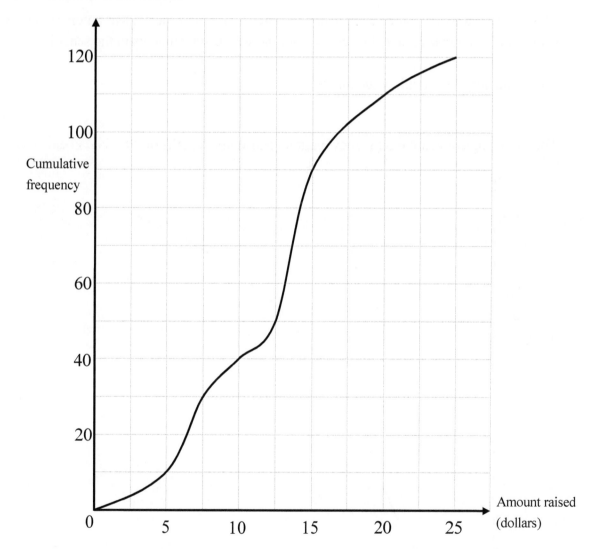

(a) (i) Write down the inter-quartile range of the amount in dollars raised.

(ii) Write down the number of students who raised between 15 dollars and 20 dollars.

[3]

The student is awarded 5 learning points per dollar for the first 15 dollars raised, and 10 learning points per dollar for each dollar they raised after the first 15 dollars.

(b) Find the number of learning points a student is awarded for raising

    (i)   15 dollars;

    (ii)   20 dollars.

[4]

(c) Find the number of students who is awarded greater than or equal to 62.5 learning points.

[3]

(d) 80 students are awarded more than k learning points. Find the value of k.

[4]

The data above was obtained from a survey. The chairman of the student union randomly chooses 120 students to take the survey from the whole 700 students in the school.

(e) Identify the sampling technique used.

[1]

2. The following cumulative frequency diagram shows the lengths of 200 fish, in cm.

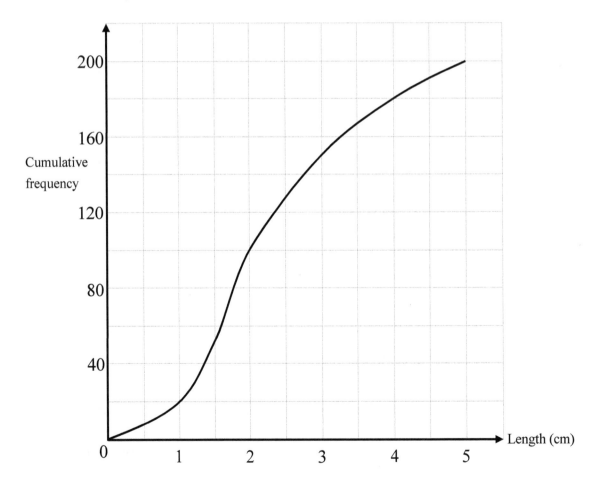

(a) (i) Write down the inter-quartile range of the length.

(ii) Write down the number of fish of length greater than 4 cm.

(iii) Find the percentage of fish of length between 1 cm and 2 cm.

(iv) 90% of the fish are longer than k cm. Find the value of k.

[8]

Fish of length greater than 3 cm are then being sold in the wet market. The price of the fish is set to be 20 dollars times the length of the fish in centimetres.

(b) Find the price of a fish of length 4.5 cm.

[2]

(c) 10% of the fish are priced higher than $\$r$. Find the value of r.

[4]

3. The following cumulative frequency diagram shows the results of a survey completed by 160 students on finding out the amount of time they spend travelling to school on a given school day.

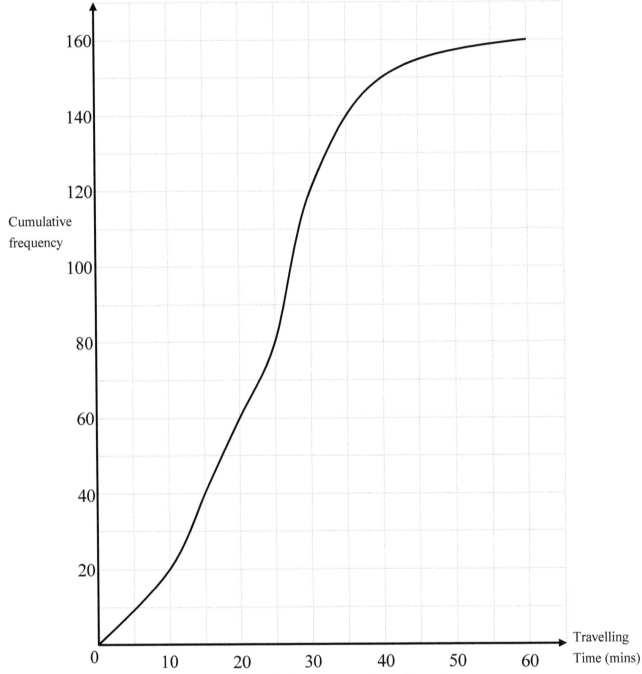

(a) Write down the median of the amount of travelling time.
[2]

(b) Write down the inter-quartile range of the amount of travelling time.
[2]

(c) Find the number of students whose travelling time is within 5 minutes of the median.
[3]

(d) 6.25% of the students spent more than k minutes to travel to school. Find the value of k.
[3]

Travelling times of greater than r minutes are considered outliers.

(e) Find the value of r.

[3]

As it was impossible to ask every student, the Head of school arranged the student names in alphabetical order and then asked every 5th student on the list.

(f) Identify the sampling technique used.

[1]

4. The following cumulative frequency diagram shows the amount of time taken by 80 secretaries to complete a presentation.

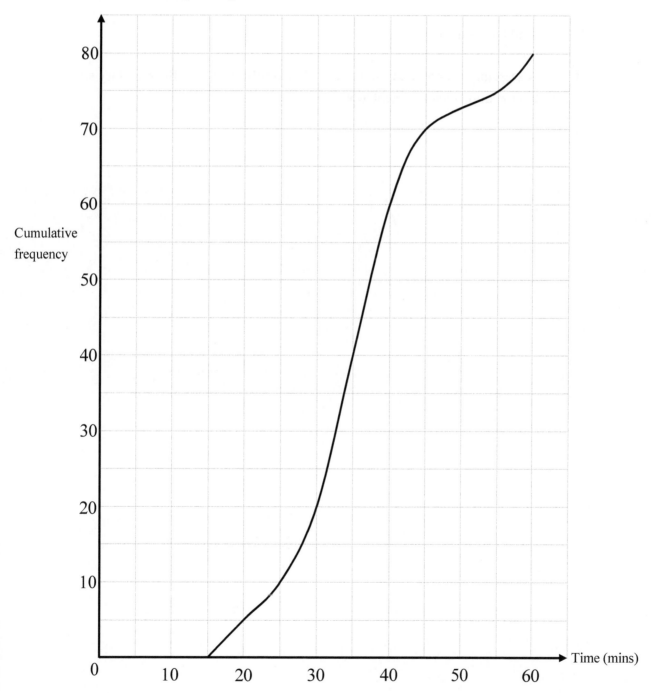

(a) Write down the median of the amount of time for presentation.
[2]

(b) Write down the inter-quartile range of the amount of time for presentation.
[2]

(c) Find the number of secretaries whose time for presentation is within 5 minutes of the upper quartile.
[3]

(d) 87.5% of the secretaries spent more than k minutes to complete a presentation. Find the value of k.
[3]

The times for presentation of greater than r minutes are considered outliers.

(e) Find the value of r.
[3]

(f) A secretary is chosen at random. Find the probability that the secretary spent more than r minutes for presentation.
[2]

Chapter 18

Probability

SUMMARY POINTs

- Terminologies:
    1. U: Universal set
    2. A: Event
    3. x: Outcome of an event
    4. $n(U)$: Total number of elements
    5. $n(A)$: Number of elements in A

- Formulae for probability:
    1. $P(A \cup B) = P(A) + P(B) - P(A \cap B)$
    2. $P(A') = 1 - P(A)$
    3. $P(A | B) = \dfrac{P(A \cap B)}{P(B)}$
    4. $P(A) = P(A \cap B) + P(A \cap B')$
    5. $P(A' \cap B') + P(A \cup B) = 1$
    6. $P(A \cup B) = P(A) + P(B)$ and $P(A \cap B) = 0$ if A and B are mutually exclusive
    7. $P(A \cap B) = P(A) \cdot P(B)$ and $P(A | B) = P(A)$ if A and B are independent

SUMMARY POINTs

✓ Venn diagram:
1. Region I: $A \cap B$
2. Region II: $A \cap B'$
3. Region III: $A' \cap B$
4. Region IV: $(A \cup B)'$

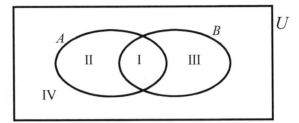

✓ Tree diagram:
1. Path I: $P(A \cap B) = pq$
2. Path I + Path III:
$$P(B) = P(A \cap B) + P(A' \cap B) = pq + (1-p)r$$

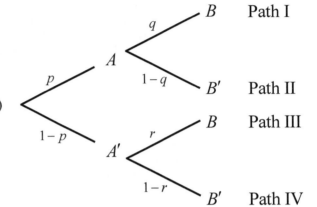

 Solutions of Chapter 18

Paper 1 – Probability with Two-Way Tables

Example

There are 25 students in a classroom, and each student plays only one sport. The table below gives their sport and gender.

	Basketball	Football	Badminton
Male	6	7	2
Female	5	3	2

(a) One student is selected at random.

    (i) Calculate the probability that the student is a female or is a basketball player.

    (ii) Given that the student selected is male, calculate the probability that the student does not play football.

[4]

(b) Two students are selected at random. Calculate the probability that neither student plays badminton.

[3]

Solution

(a) (i) The required probability

$= \dfrac{6+5+3+2}{25}$ (A1) for correct formula

$= \dfrac{16}{25}$ A1   N2

    (ii) The required probability

$= \dfrac{6+2}{6+7+2}$ (A1) for correct formula

$= \dfrac{8}{15}$ A1   N2

[4]

(b) The required probability

$$= \left(\frac{6+5+7+3}{25}\right)\left(\frac{6+5+7+3-1}{25-1}\right)$$ (A2) for correct formula

$$= \left(\frac{21}{25}\right)\left(\frac{20}{24}\right)$$

$$= \frac{7}{10}$$ A1   N3

[3]

Exercise 71

1. There are 20 students in a classroom, and each student studies only one extra language. The table below gives their choices of languages and gender.

	Spanish	French	German
Male	3	2	4
Female	3	3	5

(a) One student is selected at random.

  (i) Calculate the probability that the student is a male or studies French.

  (ii) Given that the student selected is female, calculate the probability that the student does not study Spanish.

[4]

(b) Two students are selected at random. Calculate the probability that neither student studies German.

[3]

2. There are 50 people in a house, and each person wears a uniform of different colours. The table below gives their colours of uniforms and gender.

	Red	Yellow	Green	Blue
Male	5	2	10	5
Female	10	3	5	10

(a) One person is selected at random.

  (i) Calculate the probability that the person is a female or wears a yellow uniform.

  (ii) Given that the person selected is female, calculate the probability that the person does not wear a red uniform.

[4]

(b) Two people are selected at random. Calculate the probability that both of them wear blue uniforms.

[3]

3. 25 students from three countries completed a school survey about the preferences of fruits. The table below gives their choices of fruits and nationalities.

	Apple	Orange	Banana
Japan	2	1	6
Portugal	1	5	3
Greece	4	2	1

(a) One student is selected at random.

(i) Calculate the probability that the student is not from Japan or likes apple.

(ii) Given that the student selected likes orange, calculate the probability that the student is from Portugal.

[4]

(b) Two students are selected at random. Calculate the probability that both of them are from Greece.

[3]

4. 100 students from four countries completed a questionnaire about the preferences of hobbies. The table below gives their choices of hobbies and nationalities, where a and b are positive integers.

	Hiking	Reading	Jogging
Vietnam	5	15	a
Cambodia	15	5	5
Laos	5	10	10
Thailand	5	15	b

One student is selected at random. It is given that the probability that the student is from Vietnam is $\frac{6}{25}$.

(a) Find the values of a and b.

[3]

(b) Given that the student selected does not like hiking, calculate the probability that the student is not from Cambodia.

[2]

(c) Two students are selected at random. Calculate the probability that both of them are from Thailand and like jogging.

[3]

72 Paper 1 – Probability with Venn Diagrams

Example

In a group of 25 students, 11 take arts and 14 take business. 6 students take both arts and business, as shown in the following Venn diagram. The values p and q represent numbers of students.

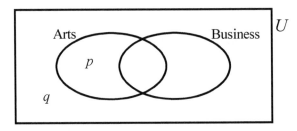

(a) Find the value of

    (i) p ;

    (ii) q .

[4]

(b) A student is selected at random. Find the probability that he takes arts but not business.

[2]

Solution

(a) (i) $p + 6 = 11$ (M1) for valid approach

           $p = 5$ A1   N2

    (ii) $5 + 14 + q = 25$ (M1) for valid approach

            $q = 6$ A1   N2

[4]

(b) The required probability

$$= \frac{5}{25}$$

                                             (M1) for valid approach

$$= \frac{1}{5}$$

                                             A1   N2

[2]

Exercise 72

1. In a group of 30 students, 21 take physics and 13 take chemistry. 9 students take both physics and chemistry, as shown in the following Venn diagram. The values a and b represent numbers of students.

   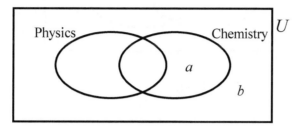

   (a) Find the value of

   (i) a;

   (ii) b.

   [4]

   (b) A student is selected at random. Find the probability that he takes chemistry only.

   [2]

2. In a group of 40 students, 17 are in the choir and 15 are in the school band. 10 students are neither in the choir nor in the band, as shown in the following Venn diagram. The values h and k represent numbers of students.

   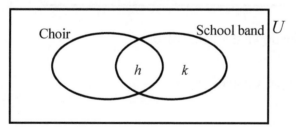

   (a) Find the value of

   (i) h;

   (ii) k.

   [4]

   (b) A student is selected at random. Find the probability that he is in both the choir and the school band.

   [2]

3. The following Venn diagram shows the events A and B, where $P(B) = 0.6$, $P(A \cup B) = 0.9$ and $P(A \cap B) = 0.4$. The values p and q are probabilities.

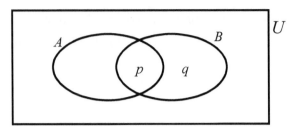

(a) (i) Write down the value of p.

(ii) Find the value of q.

[3]

(b) Find $P(A)$.

[3]

4. The following Venn diagram shows the events A and B, where $P(A) = 0.6$, $P(B) = 0.2$ and $P(A' \cap B') = 0.3$. The values a and b are probabilities.

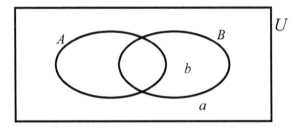

(a) (i) Write down the value of a.

(ii) Find the value of b.

[3]

(b) Find $P(A \cap B)$.

[3]

73 Paper 1 – Probability with Tree Diagrams

Example

A bag contains 5 black balls and 2 white balls. Two balls are selected at random without replacement.

(a) Complete the following tree diagram.

[3]

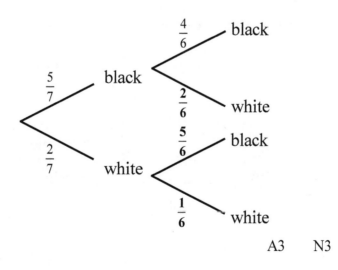

(b) Find the probability that exactly one of the selected balls is black.

[3]

Solution

(a)

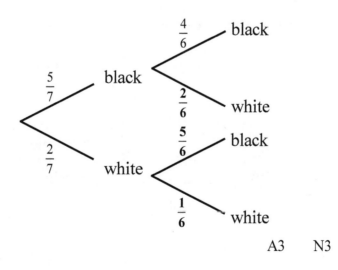

A3   N3

[3]

(b) The required probability

$= \left(\dfrac{5}{7}\right)\left(\dfrac{2}{6}\right) + \left(\dfrac{2}{7}\right)\left(\dfrac{5}{6}\right)$ (M1)(A1) for correct formula

$= \dfrac{10}{21}$ A1   N3

[3]

Exercise 73

1. A bag contains 3 red balls and 5 blue balls. Two balls are selected at random without replacement.

   (a) Complete the following tree diagram.

   [3]

   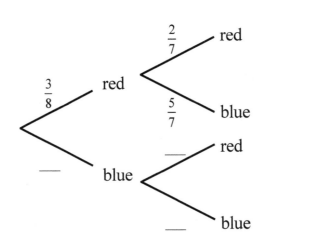

   (b) Find the probability that exactly one of the selected balls is blue.

   [3]

2. A bag contains 4 yellow marbles and 5 purple marbles. Two marbles are selected at random without replacement.

   (a) Complete the following tree diagram.

   [3]

   (b) Find the probability that at least one of the selected marbles is purple.

   [3]

3. The diagram shows the probabilities for events A and B, with $P(A') = x$.

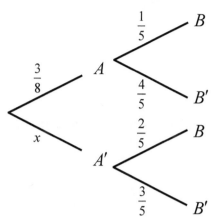

(a) Write down the value of x.

[1]

(b) Find $P(B)$.

[3]

(c) Find $P(A|B)$.

[3]

4. The diagram shows the probabilities for events A and B.

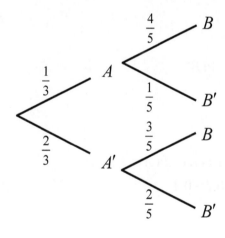

(a) Write down the value of $P(B|A')$.

[1]

(b) Find $P(B)$.

[3]

(c) Find $P(A'|B)$.

[3]

Your Practice Set – Applications and Interpretation for IBDP Mathematics

 Paper 1 – Probability Formulae

Example

Events A and B are independent with $P(A \cap B) = P(A' \cap B) = 0.1$.

(a) Find $P(B)$. [2]

(b) Find $P(A)$. [2]

(c) Find $P(A \cup B)$. [2]

Solution

(a) $P(B) = P(A \cap B) + P(A' \cap B)$ (M1) for valid approach
$P(B) = 0.1 + 0.1$
$P(B) = 0.2$ A1 N2 [2]

(b) $P(A \cap B) = P(A) \times P(B)$
$0.1 = P(A) \times 0.2$ (A1) for substitution
$P(A) = 0.5$ A1 N2 [2]

(c) $P(A \cup B) = P(A) + P(B) - P(A \cap B)$
$P(A \cup B) = 0.5 + 0.2 - 0.1$ (A1) for substitution
$P(A \cup B) = 0.6$ A1 N2 [2]

Exercise 74

1. Events A and B are independent with $P(A \cap B) = 0.08$ and $P(A \cap B') = 0.12$.

   (a) Find $P(A)$. [2]

   (b) Find $P(B)$. [2]

   (c) Find $P(A \cup B)$. [2]

2. Events A and B are independent with $P(A' \cap B) = 0.15$ and $P(B) = 0.3$.

   (a) Find $P(A \cap B)$.

   [2]

   (b) Find $P(A)$.

   [2]

   (c) Find $P(A \cup B)$.

   [2]

3. Events A and B are independent with $P(A \cap B') = 0.28$ and $P(A) = 0.4$.

   (a) Find $P(A \cap B)$.

   [2]

   (b) Find $P(A \cup B)$.

   [4]

4. Events A and B are independent with $P(A \cap B) = 0.21$ and $P(A) = 0.7$.

   (a) Find $P(B)$.

   [2]

   (b) Show that $P(A \cup B) = 0.79$.

   [2]

   (c) Find $P(A' \cap B')$.

   [3]

Your Practice Set – Applications and Interpretation for IBDP Mathematics

Paper 1 – Setting Up Probability Equations

Example

Let A and B be independent events, with $P(A) = k$ and $P(B) = 2k^2$, where $0 < k < 0.5$.

(a) Write down an expression for $P(A \cap B)$ in terms of k.

[2]

(b) Given that $P(A \cap B) = 0.054$, find k.

[2]

(c) Find $P(A' \cap B)$.

[2]

(d) Find $P(A' | B)$.

[2]

Solution

(a) $P(A \cap B) = P(A) \times P(B)$

$P(A \cap B) = k \times 2k^2$ (A1) for substitution

$P(A \cap B) = 2k^3$ A1   N2

[2]

(b) $2k^3 = 0.054$ (A1) for correct equation

$k^3 = 0.027$

$k = 0.3$ A1   N2

[2]

(c) $P(B) = P(A \cap B) + P(A' \cap B)$

$2(0.3)^2 = 0.054 + P(A' \cap B)$ (A1) for substitution

$P(A' \cap B) = 0.126$ A1   N2

[2]

(d) $P(A' | B) = \dfrac{P(A' \cap B)}{P(B)}$

$P(A' | B) = \dfrac{0.126}{2(0.3)^2}$ (A1) for substitution

$P(A' | B) = 0.7$ A1   N2

[2]

Exercise 75

1. Let C and D be independent events, with $P(C) = 2k^2$ and $P(D) = 3k^2$, where $0 < k < 0.5$.

   (a) Write down an expression for $P(C \cap D)$ in terms of k.

   [2]

   (b) Given that $P(C \cap D) = 0.0096$, find k.

   [2]

   (c) Find $P(C \cap D')$.

   [2]

   (d) Find $P(D'|C)$.

   [2]

2. Let E and F be independent events, with $P(E) = 4k^3$ and $P(F) = k$, where $0 < k < \frac{1}{2}$.

   (a) Write down an expression for $P(E \cap F)$ in terms of k.

   [2]

   (b) Given that $P(E \cap F) = \frac{1}{2500}$, find k.

   [2]

   (c) Find $P(E \cup F)$.

   [2]

3. Let A and B be independent events, with $P(A) = 2k$ and $P(B) = 1.5 P(A)$, where $0 < k < 0.5$. It is given that $P(A \cup B) = 6k - 1$.

   (a) Show that $P(A \cap B) = 6k^2$.

   [2]

   (b) Using $P(A \cup B) = P(A) + P(B) - P(A \cap B)$, show that $6k^2 + k - 1 = 0$.

   [2]

   (c) Hence, find the value of k.

   [2]

   (d) Find $P(B|A)$.

   [2]

Your Practice Set – Applications and Interpretation for IBDP Mathematics

4. Consider the independent events A and B. Given that $P(B) = 3P(A)$, and $P(A \cup B) = 0.93$, find $P(B)$.

    (a) Let $x = P(A)$. Show that $300x^2 - 400x + 93 = 0$.

[4]

    (b) Hence, find $P(B)$.

[3]

76 Paper 2 – Miscellaneous Problems

Example

In a school, students are required to learn at least one language, Japanese or Korean. It is known that 80 % of the students learn Japanese, and 55 % learn Korean.

(a) Find the percentage of students who learn **both** Japanese and Korean.

[2]

(b) Find the percentage of students who learn Japanese, but not Korean.

[2]

At this school, 58 % of the students are female, and 75 % of the female learn Japanese. A student is chosen at random.

(c) (i) Find the probability that the student is a female and she learns Japanese.

(ii) Find the probability that the student is a male and he learns Japanese.

(iii) Find the probability that the student is a male.

(iv) A male is chosen at random. Find the probability that he learns Japanese.

[9]

Solution

(a) $P(J \cup K) = P(J) + P(K) - P(J \cap K)$
$1 = 0.8 + 0.55 - P(J \cap K)$ (A1) for substitution
$P(J \cap K) = 0.35$
Thus, the required percentage is 35%.   A1   N2

[2]

(b) $P(J \cap K) + P(J \cap K') = P(J)$ (M1) for valid approach
$0.35 + P(J \cap K') = 0.8$
$P(J \cap K') = 0.45$
Thus, the required percentage is 45%.   A1   N2

[2]

(c) (i) The required probability
$= P(F \cap J)$
$= P(J|F) \times P(F)$ (M1) for valid approach
$= 0.75 \times 0.58$ (A1) for substitution
$= 0.435$ A1   N3

(ii) $P(F \cap J) + P(F' \cap J) = P(J)$ (M1) for valid approach
$0.435 + P(F' \cap J) = 0.8$
$P(F' \cap J) = 0.365$
Thus, the required probability is 0.365.   A1   N2

(iii) $P(F) + P(F') = 1$
$0.58 + P(F') = 1$ (A1) for substitution
$P(F') = 0.42$
Thus, the required probability is 0.42.   A1   N2

(iv) The required probability
$= P(J|F')$
$= \dfrac{P(F' \cap J)}{P(F')}$
$= \dfrac{0.365}{0.42}$ (A1) for substitution
$= 0.869047619$
$= 0.869$ A1   N2

[9]

Exercise 76

1. Consider the events A and B. It is given that $P(A) = 0.4$, $P(B) = 0.65$ and $P(A \cup B) = 1$.

   (a) Find $P(A \cap B)$.

   [2]

   (b) Find $P(A' \cap B)$.

   [2]

   Consider another event C. It is also given that $P(C) = 0.7$ and $P(A | C) = 0.78$.

   (c) (i) Find $P(A \cap C)$.

   (ii) Find $P(A' \cap C)$.

   (iii) Find $P(A')$.

   (iv) Hence, find $P(C | A')$.

   [9]

2. In a school, students are required to take at least one subject from art and technology. It is known that 55 % of the students take art, and 70 % take technology.

   (a) Find the percentage of students who take **both** art and technology.

   [2]

   (b) Find the percentage of students who take one subject only.

   [2]

   At this school, 63 % of the students are male, and 72 % of the male take art. A student is chosen at random.

   (c) (i) Find the probability that the student chosen is a male art student.

   (ii) Find the probability that the student chosen is a female art student.

   (iii) Find the probability that a female student is chosen.

   (iv) Given that a female student is chosen, find the probability that she takes art.

   [9]

Your Practice Set – Applications and Interpretation for IBDP Mathematics

3. In a class of 100 boys, 85 boys play football and 45 boys play rugby. Each boy must play at least one sport from football and rugby.

   (a) Find the percentage of boys who play both sports.

   [2]

   (b) Find the percentage of boys who play rugby only.

   [2]

   (c) Find the percentage of boys who play only one sport.

   [2]

   (d) One boy is selected at random.

   (i) Given that the boy selected plays football, find the probability that he plays rugby also.

   (ii) Given that the boy selected plays only one sport, find the probability that he plays rugby.

   [4]

   It is also given that 60% of the boys are taller than 180 cm and 90% of these boys play football.

   (e) (i) Write down the percentage of boys who is taller than 180 cm and play football.

   (ii) Hence, find the percentage of boys who is not taller than 180 cm and play football.

   [3]

4. A survey of the reading habits of a group of students revealed that 35% read free newspapers, 50% read paid newspapers and 25% do not read newspapers at all.

   (a) Find the percentage of students who read both types of newspapers.

   [2]

   (b) Find the percentage of students who only read paid newspapers.

   [2]

   (c) Write down the percentage of students who read newspapers.

   [1]

   (d) One student is selected at random.

   (i) Given that the student selected reads newspapers, find the probability that he reads paid newspapers only.

   (ii) Given that the student selected reads paid newspapers, find the probability that he reads free newspapers also.

   [4]

It is also given that 40% of the students wear glasses and 95% of these students read paid newspapers.

(e) (i) Write down the percentage of students who wear glasses and read paid newspapers.

(ii) Hence, find the percentage of students who do not wear glasses and read paid newspapers.

[3]

Chapter 19

Discrete Probability Distributions

SUMMARY POINTs

✓ Properties of a discrete random variable X:

X	x_1	x_2	$\cdots$	x_n
$P(X=x)$	$P(X=x_1)$	$P(X=x_2)$	$\cdots$	$P(X=x_n)$

1. $P(X = x_1) + P(X = x_2) + \cdots + P(X = x_n) = 1$
2. $E(X) = x_1 P(X = x_1) + x_2 P(X = x_2) + \cdots + x_n P(X = x_n)$: Expected value of X
3. $E(X) = 0$ if a fair game is considered

Solutions of Chapter 19

Paper 1 – Expectation of a Variable

Example

The following table shows the probability distribution of a discrete random variable X.

x	0	2	4	6
$P(X = x)$	0.4	0.1	$2k$	$3k$

(a) Find the value of k.

[2]

(b) Find $E(X)$.

[3]

Solution

(a) $P(X=0)+P(X=2)+P(X=4)+P(X=6)=1$ (A1) for correct formula
$0.4+0.1+2k+3k=1$
$5k=0.5$
$k=0.1$ A1   N2

[2]

(b) $E(X) = (0)(0.4)+(2)(0.1)+(4)(2k)+(6)(3k)$ (A1) for correct formula
$E(X) = (0)(0.4)+(2)(0.1)+(4)(0.2)+(6)(0.3)$ (A1) for substitution
$E(X) = 2.8$ A1   N3

[3]

Exercise 77

1. The following table shows the probability distribution of a discrete random variable X.

x	0	1	2	3
$P(X = x)$	$9k$	k	0.1	0.4

(a) Find the value of k.

[2]

(b) Find $E(X)$.

[3]

Your Practice Set – Applications and Interpretation for IBDP Mathematics

2. The following table shows the probability distribution of a discrete random variable X.

x	0	20	40	60
$P(X=x)$	$\dfrac{1}{10}$	$\dfrac{1}{5}$	$\dfrac{2}{5}$	k

   (a) Find the value of k.

   [2]

   (b) Find $E(X)$.

   [3]

3. The following table shows the probability distribution of a discrete random variable X. It is given that $k > 0$.

x	1	2	k
$P(X=x)$	$\dfrac{1}{14}$	$\dfrac{4}{14}$	$\dfrac{k^2}{14}$

   (a) Find the value of k.

   [3]

   (b) Find $E(X)$.

   [3]

   The observation of X is recorded for two times. Let Y be the sum of these two observations.

   (c) Find $P(Y=3)$.

   [2]

4. The following table shows the probability distribution of a discrete random variable X.

x	k	$k+1$	$k+2$	8
$P(X=x)$	$\dfrac{k}{2}$	$\dfrac{1}{8}$	$\dfrac{k}{4}$	$\dfrac{1}{8}$

   (a) Find the value of k.

   [3]

   (b) Find $E(X)$.

   [3]

   The observation of X is recorded for two times. Let Y be the product of these two observations.

   (c) Find $P(Y=2)$.

   [2]

78. Paper 1 – Conditional Probabilities in a Table

Example

A discrete random variable X has the following probability distribution.

x	1	2	3	4
$P(X=x)$	$5k^2$	$\dfrac{1}{2}k$	$\dfrac{3}{2}k$	$3k^2$

(a) Find the value of k.

[3]

(b) Find $P(X=4 \mid X>2)$.

[3]

Solution

(a) $P(X=1)+P(X=2)+P(X=3)+P(X=4)=1$

$5k^2+\dfrac{1}{2}k+\dfrac{3}{2}k+3k^2=1$ (A1) for substitution

$8k^2+2k-1=0$

$(2k+1)(4k-1)=0$ (A1) for factorization

$k=-\dfrac{1}{2}$ (Rejected) or $k=\dfrac{1}{4}$ A1   N3

[3]

(b) $P(X=4 \mid X>2)=\dfrac{P(X=4 \cap X>2)}{P(X>2)}$

$P(X=4 \mid X>2)=\dfrac{P(X=4)}{P(X=3)+P(X=4)}$ (M1) for valid approach

$P(X=4 \mid X>2)=\dfrac{3\left(\dfrac{1}{4}\right)^2}{\dfrac{3}{2}\left(\dfrac{1}{4}\right)+3\left(\dfrac{1}{4}\right)^2}$ (A1) for substitution

$P(X=4 \mid X>2)=\dfrac{1}{3}$ A1   N3

[3]

Your Practice Set – Applications and Interpretation for IBDP Mathematics

Exercise 78

1. A discrete random variable X has the following probability distribution.

x	4	8	12
$P(X = x)$	$10k^2$	k	$20k^2$

   (a) Find the value of k. [3]

   (b) Find $P(X = 12 \mid X > 6)$. [3]

2. A discrete random variable X has the following probability distribution.

x	12	24	30	36
$P(X = x)$	k	$7k^2$	$8k^2$	k

   (a) Find the value of k. [3]

   (b) Find $P(X = 24 \mid X > 20)$. [3]

3. A discrete random variable X has the following probability distribution.

x	7	14	21	28	35
$P(X = x)$	k	$3k$	$10k^2$	$6k^2$	$5k^2$

   (a) Find the value of k. [3]

   (b) Find $P(X < 15 \mid X < 25)$. [3]

4. A discrete random variable X has the following probability distribution.

x	0	1	2	3	4	5
$P(X = x)$	k^2	k	$4k^2$	$8k^2$	$4k$	k^2

   (a) Find the value of k. [3]

   (b) Find $P(2 < X \leq 4 \mid 1 < X \leq 4)$. [3]

Paper 1 – Unknown Probabilities with Given $E(X)$

Example

The random variable X has the following probability distribution.

x	2	4	6
$P(X=x)$	a	b	0.25

(a) Find the first equation connecting a and b, giving the answer in the form $pa + qb = r$.

[2]

It is given that $E(X) = 3.4$.

(b) Find the second equation connecting a and b, giving the answer in the form $pa + qb = r$.

[2]

(c) Hence, write down the values of a and b.

[2]

Solution

(a) $P(X=2) + P(X=4) + P(X=6) = 1$ (A1) for correct formula
$a + b + 0.25 = 1$
$a + b = 0.75$ A1   N2

[2]

(b) $E(X) = 3.4$
$2a + 4b + (6)(0.25) = 3.4$ (A1) for correct formula
$2a + 4b = 1.9$ A1   N2

[2]

(c) $a = 0.55$, $b = 0.2$ A2   N2

[2]

Your Practice Set – Applications and Interpretation for IBDP Mathematics

Exercise 79

1. The random variable X has the following probability distribution.

x	1	2	3	4
$P(X=x)$	0.2	0.3	a	b

   (a) Find the first equation connecting a and b, giving the answer in the form $pa+qb=r$.

   [2]

   It is given that $E(X) = 2.62$.

   (b) Find the second equation connecting a and b, giving the answer in the form $pa+qb=r$.

   [2]

   (c) Hence, write down the values of a and b.

   [2]

2. The random variable X has the following probability distribution.

x	20	30	40	50
$P(X=x)$	0.1	a	b	0.1

   It is given that $P(X<45) = 0.9$.

   (a) Find the first equation connecting a and b, giving the answer in the form $pa+qb=r$.

   [2]

   It is given that $E(X) = 33$.

   (b) Find the second equation connecting a and b, giving the answer in the form $pa+qb=r$.

   [2]

   (c) Hence, write down the values of a and b.

   [2]

3. The random variable X has the following probability distribution.

x	0	10	20	30
$P(X=x)$	0.1	a	b	c

It is given that $P(X<15)=0.5$.

(a) Find the value of a.

[2]

It is also given that $E(X)=16$.

(b) Find the values of b and c.

[4]

The observation of X is recorded for two times. Let Y be the sum of these two observations.

(c) Find $P(Y=50)$.

[2]

4. The random variable X has the following probability distribution.

x	0	3	6	9
$P(X=x)$	0.4	a	b	c

It is given that $P(2<X<7)=0.3$.

(a) Find the value of c.

[2]

It is also given that $E(X)=4.2$.

(b) Find the values of a and b.

[4]

The observation of X is recorded for two times. Let Y be the product of these two observations.

(c) Find $P(Y=36)$.

[2]

80 Paper 2 – Problems Involving Tree Diagrams

Example

Petr travels to school on a train. On any day, the probability that Petr will miss the train is 0.2.

If he misses the train, the probability that he will be late for school is 0.9. If he does not miss the train, the probability that he will be late is 0.25.

Let A be the event "he misses the train" and B the event "he is late for school".

The information above is shown on the following tree diagram.

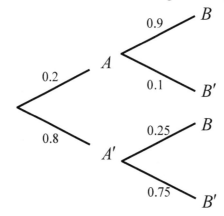

(a) Find

    (i) $P(A \cap B)$;

    (ii) $P(B)$.

[4]

(b) Find the probability that

    (i)   Petr does not miss the train and is not late for school;

    (ii)   Petr does not miss the train, given that he is late for school.

[5]

The cost for each day that Petr catches the train is 5 dollars. Petr goes to school on Monday and Friday. Let X be the total transportation cost.

(c) **Copy** and complete the probability distribution table.

[3]

X	0	5	10
$P(X = x)$			

(d) Find the expected cost for Petr for both days.

[2]

Solution

(a) (i) $P(A \cap B) = (0.2)(0.9)$ (A1) for substitution
       $P(A \cap B) = 0.18$ A1   N2

(ii) $P(B) = P(A \cap B) + P(A' \cap B)$ (M1) for valid approach
       $P(B) = 0.18 + (0.8)(0.25)$
       $P(B) = 0.38$ A1   N2

[4]

(b) (i)   The required probability
       $= P(A' \cap B')$
       $= (0.8)(0.75)$ (A1) for substitution
       $= 0.6$ A1   N2

(ii)   The required probability
       $= P(A' \mid B)$ (M1) for valid approach
       $= \dfrac{P(A' \cap B)}{P(B)}$
       $= \dfrac{(0.8)(0.25)}{0.38}$ (A1) for substitution
       $= \dfrac{10}{19}$ A1   N3

[5]

(c)

X	0	5	10
$P(X = x)$	0.04	0.32	0.64

                                                   A3   N3

[3]

(d)   The expected cost
       $= (0)(0.04) + (5)(0.32) + (10)(0.64)$ (A1) for substitution
       $= \$8$ A1   N2

[2]

Exercise 80

1. A **five-sided fair** die has three blue faces and two red faces. The die is rolled for two times.

   Let F be the event a blue face lands down in the first roll, and S be the event a blue face lands down in the second roll.

   The information above is shown on the following tree diagram.

   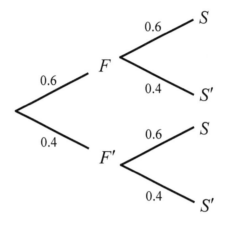

   (a) Find

       (i) $P(F \cap S)$;

       (ii) $P(S)$.

   [4]

   (b) Find the probability that

       (i)   a red face lands down for two times;

       (ii)   a blue face lands down in the first roll, given that a red face lands down in the second roll.

   [5]

   Rossi plays a game where he rolls the die for two times. If a blue face lands down, he scores 4. If the red face lands down, he scores 1. Let X be the total score obtained.

   (c) **Copy** and complete the probability distribution table.

   [3]

X	2	5	8
$P(X = x)$			

   (d) Calculate the expected value of X.

   [2]

2. Sandy and Katie both wish to go to the shopping mall but one of them has to stay home for baby-sitting. The probability that Sandy goes to the shopping mall is 0.2. Exactly one of them will go. If Sandy goes to the shopping mall the probability that she is late back home is 0.7. If Katie goes to the shopping mall the probability that she is late back home is 0.4.

Let S be the event that Sandy goes to the shopping mall, and L be the event that the person who goes to the shopping mall arrives home late. The information above is shown on the following tree diagram.

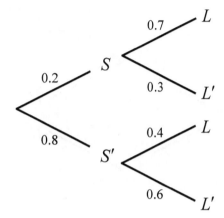

(a) Find

   (i) $P(S \cap L')$;

   (ii) $P(L')$.

[4]

(b) Find the probability that

   (i) Katie goes to the shopping mall and she is not late;

   (ii) Sandy goes to the shopping mall, given that the person who goes to the shopping mall arrives home late.

[5]

The attendance of Sandy is recorded for two particular days when she goes to the shopping mall. If one lateness is marked, she will be penalized by a $10 fee reduction. If two latenesses are marked, she will be penalized by a $25 fee reduction. Let X be the total amount of fee reduction for Sandy.

(c) **Copy** and complete the probability distribution table.

[3]

X	0	10	25
$P(X = x)$			

(d) Calculate the expected value of X.

[2]

3. Thierry and Jake share a flat. Thierry cooks dinner five nights out of ten. If Thierry does not cook dinner then Jake does. If Thierry cooks dinner the probability that they have lasagna is $\frac{9}{10}$. If Jake cooks dinner the probability that they have lasagna is $\frac{3}{10}$.

Let T be the event that Thierry cooks dinner, and L be the event that they have lasagna in their dinner.

The information above is shown on the following tree diagram.

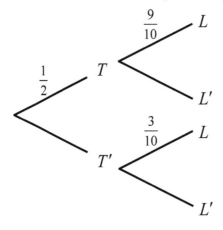

(a) Find

   (i) $P(T' \cap L')$;

   (ii) $P(L')$.

[4]

(b) Find the probability that

   (i) Thierry cooks dinner and they do not have lasagna;

   (ii) they do not have lasagna, given that Jake cooks dinner.

[5]

The cost for making lasagna for once is 125 dollars. Consider the expenditure on making lasagna on a particular three days. Let X be the total amount of expenditure.

(c) **Copy** and complete the probability distribution table.

[3]

X	0	125	250	375
$P(X = x)$			$\frac{54}{125}$	

(d) Find the expected expenditure.

[2]

4. On a shelf there are two tins, one red and one blue. The red tin contains two orange candies and eight apple candies, and the blue tin contains six orange candies and four apple candies. Oliver randomly chooses either the red or the blue tin and randomly selects a candy.

Let R be the event that the red tin is chosen, and A be the event that an apple candy is selected. The information above is shown on the following tree diagram.

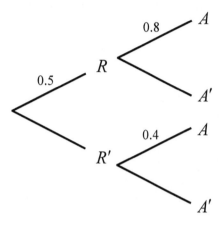

(a) Find

  (i) $P(R' \cap A)$;

  (ii) $P(A)$.

[4]

(b) Find the probability that

  (i) an orange candy is selected from the red tin;

  (ii) he selected a candy from the red tin, given that the candy selected is an apple candy.

[5]

Oliver designed a game by using the above two tins. If an orange candy is chosen, then the participant will get 4 tokens. If an apple candy is chosen, no tokens would be awarded. Let X be the total number of tokens of three different participants of the game.

(c) **Copy** and complete the probability distribution table.

[3]

X	0	4	8	12
$P(X = x)$			$\dfrac{36}{125}$	

(d) Find the expected expenditure.

[2]

Your Practice Set – Applications and Interpretation for IBDP Mathematics

 Paper 2 – Game Problems

Example

In a game, two standard six-sided dice are tossed. Let X be the sum of the scores on the two dice.

(a)　Find

　　(i)　$P(X = 10)$;

　　(ii)　$P(X > 10)$;

　　(iii)　$P(X = 11 \mid X > 9)$.

[7]

(b)　Jenna plays a game where she tosses two dice.

　　If the sum is 10, she wins 2 points.
　　If the sum is greater than 10, she wins 1 point.
　　If the sum is less than 10, she loses k points.

　　(i)　Write down the probability that the sum is less than 10.

　　(ii)　Hence, find the value of k for which Jenna's expected number of points is zero.

[5]

Solution

(a)　(i)　There are 3 ways such that $X = 10$　　(A1) for correct value

$$P(X = 10) = \frac{3}{36}$$

$$P(X = 10) = \frac{1}{12}$$　　A1　N2

　　(ii)　There are 3 ways such that $X > 10$　　(A1) for correct value

$$P(X > 10) = \frac{3}{36}$$

$$P(X > 10) = \frac{1}{12}$$　　A1　N2

(iii) $P(X=11 | X>9) = \dfrac{P(X=11 \cap X>9)}{P(X>9)}$ (M1) for valid approach

$P(X=11 | X>9) = \dfrac{P(X=11)}{P(X>9)}$

$P(X=11 | X>9) = \dfrac{\frac{2}{36}}{\frac{1}{12}+\frac{1}{12}}$ (A1) for substitution

$P(X=11 | X>9) = \dfrac{1}{3}$ A1   N3

[7]

(b) (i) $\dfrac{5}{6}$ A1   N1

(ii) $E(X) = 0$
$(2)P(X=10) + (1)P(X>10)$
$+(-k)P(X<10) = 0$ (M1)(A1) for correct formula

$(2)\left(\dfrac{1}{12}\right) + (1)\left(\dfrac{1}{12}\right) + (-k)\left(\dfrac{5}{6}\right) = 0$ (A1) for substitution

$2 + 1 - 10k = 0$

$k = \dfrac{3}{10}$ A1   N4

[5]

Exercise 81

1. In a game, two standard six-sided dice are tossed. Let X be the sum of the scores on the two dice.

   (a) Find

   (i) $P(X = 5)$;

   (ii) $P(X < 5)$;

   (iii) $P(X = 4 | X < 6)$.

   [7]

(b) Macy plays a game where she tosses two dice.

If the sum is 5, she wins 3 points.
If the sum is less than 5, she wins 2 points.
If the sum is greater than 5, she loses k points.

(i) Write down the probability that the sum is greater than 5.

(ii) Find the value of k for which Macy's expected number of points is zero.

[5]

2. In a game, two standard six-sided dice are tossed. Let X be the sum of the scores on the two dice.

(a) Find

(i) $P(X = 8)$;

(ii) $P(X > 8)$;

(iii) $P(X > 9 \mid X > 8)$.

[7]

(b) Rick plays a game where he tosses two dice.

If the sum is 8, he wins 5 points.
If the sum is greater than 8, he wins k points.
If the sum is less than 8, he loses 1 point.

(i) Write down the probability that the sum is less than 8.

(ii) Find the value of k for which Rick's expected number of points is one.

[5]

3. In a game, there are two different boxes containing different numbered cards. The first box contains cards numbered 1, 2 and 3. The second box contains cards numbered 10, 20 and 30. Two cards are drawn at random, one from each box. Let X be the sum of the numbers on the two cards.

(a) Find

(i) $P(X = 21)$;

(ii) $P(X > 21)$;

(iii) $P(30 < X < 33 \mid X > 21)$.

[7]

(b) Stephen plays a game where he draws two cards at random, one from each box.

If the sum is 21, he wins $3k$ points.
If the sum is greater than 21, he wins k points.
If the sum is less than 21, no points would be awarded

Find the value of k for which Stephen's expected number of points is eight.

[6]

4. In a game, there are two different boxes containing different numbered cards. The first box contains cards numbered 2, 3 and 5. The second box contains cards numbered 2, 7 and 11. Two cards are drawn at random, one from each box. Let X be the product of the numbers on the two cards.

(a) Find

(i) $P(X = 33)$;

(ii) $P(X \geq 35)$;

(iii) $P(X < 22 \mid X < 33)$.

[7]

(b) Jessie plays a game where she draws two cards at random, one from each box.

If the product is 33, she wins $4k$ points.
If the product is greater than 33, she wins $3k$ points.
If the product is less than 33, she loses $2k$ points.

Find the value of k for which Jessie's expected number of points is -16.

[6]

Chapter 20

Binomial Distribution

SUMMARY POINTs

✓ Properties of a random variable $X \sim B(n, p)$ following binomial distribution:
1. Only two outcomes from every independent trial (Success and failure)
2. n: Number of trials
3. p: Probability of success
4. X: Number of successes in n trials

✓ Formulae for binomial distribution:
1. $P(X = r) = \binom{n}{r} p^r (1-p)^{n-r}$ for $0 \leq r \leq n$
2. $E(X) = np$: Expected value of X
3. $\text{Var}(X) = np(1-p)$: Variance of X
4. $\sqrt{np(1-p)}$: Standard deviation of X
5. $P(X \leq r) = P(X < r+1) = 1 - P(X \geq r+1)$

Solutions of Chapter 20

82 Paper 1 – Expected Value and Variance

Example

A biased coin is tossed for twelve times. The probability of getting a head is 0.6.

(a) Find the expected number of heads landed.
[2]

(b) Find the variance of the number of heads landed.
[2]

(c) Find the probability that the number of heads landed is equal to the number of tails landed.
[2]

Solution

(a) The expected number
$= (12)(0.6)$ (A1) for substitution
$= 7.2$ A1   N2
[2]

(b) The variance
$= (12)(0.6)(1-0.6)$ (A1) for substitution
$= 2.88$ A1   N2
[2]

(c) The required probability
$= \binom{12}{6}(0.6)^6(1-0.6)^{12-6}$ (A1) for substitution
$= 0.176579149$
$= 0.177$ A1   N2
[2]

Your Practice Set – Applications and Interpretation for IBDP Mathematics

Exercise 82

1. A fair eight-faced die with numbered faces 1, 2, 3, 4, 5, 6, 7 and 8 is tossed for ten times.

    (a) Find the expected number of multiples of 4 landed.
    [2]

    (b) Find the variance of the number of multiples of 4 landed.
    [2]

    (c) Find the probability that the number of multiples of 4 landed is 3.
    [2]

2. A random variable X follows a binomial distribution such that $X \sim B(n, 0.6)$, where n is a positive integer. It is given that the expected value of X is 18.

    (a) Find the value of n.
    [2]

    (b) Find the value of $\mathrm{Var}(X)$.
    [2]

    (c) Find the value of $P(X = 19)$.
    [2]

3. A game spinner is produced such that the probability of getting a yellow color is landed is p, where $p > 0.5$. The spinner is tossed for 20 times. It is given that the variance of the number of yellow color landed is 3.2.

    (a) Find the value of p.
    [3]

    (b) Find the expected number of yellow color landed.
    [2]

    (c) Find the probability of getting a yellow color for 17 times.
    [2]

4. Paul lives in a city. He believes that a proportion p of all n citizens in the city will vote for him in an election. It is given that the mean and the variance of the number of citizens voting for Paul are 1800 and 990 respectively.

    (a) Find the value of p.
    [3]

    (b) Find the value of n.
    [2]

    (c) Write down the probability that only half of all citizens vote for him, giving the answer in the form $a \times 10^k$, where $1 \leq a < 10$ and $k \in \mathbb{Z}$, and correct a to 4 decimal places.
    [2]

83. Paper 1 – Probabilities in a Binomial Distribution

Example

A box holds 100 eggs. The probability that an egg is white is 0.04. Let X be the number of white eggs in the box.

(a) Find the expected number of white eggs in the box.
[2]

(b) Find the probability that there are exactly 13 white eggs in the box.
[2]

(c) Find the probability that there are at least 8 white eggs in the box.
[3]

Solution

(a) $E(X) = (100)(0.04)$ (A1) for substitution
$E(X) = 4$ A1   N2
[2]

(b) $P(X = 13)$
$= \binom{100}{13}(0.04)^{13}(1-0.04)^{100-13}$ (A1) for substitution
$= 0.000136861$
$= 0.000137$ A1   N2
[2]

(c) $P(X \geq 8)$
$= 1 - P(X \leq 7)$ (M1) for valid approach
$= 1 - 0.952487949$ (A1) for correct value
$= 0.047512051$
$= 0.0475$ A1   N3
[3]

Your Practice Set – Applications and Interpretation for IBDP Mathematics

Exercise 83

1. A box holds 80 apples. The probability that an apple is rotten is 0.06. Let X be the number of rotten apples in the box.

    (a) Find the expected number of rotten apples in the box.
    [2]

    (b) Find the probability that there are exactly 10 rotten apples in the box.
    [2]

    (c) Find the probability that there are at least 15 rotten apples in the box.
    [3]

2. In a school the probability that a student is left-handed is 0.12. A sample of 135 students is randomly selected from the school. Let X be the number of left-handed students in the sample.

    (a) Find the expected number of left-handed students in this sample.
    [2]

    (b) Find the probability that exactly 20 students are left-handed.
    [2]

    (c) Find the probability that more than 16 students are left-handed.
    [3]

3. A factory makes lamps. The probability that a lamp is defective is 0.02. The factory tests a random sample of 50 lamps. Let X be the number of defective lamps in the sample.

    (a) Find the mean number of defective lamps in the sample.
    [2]

    (b) Find the probability that there are exactly nine defective lamps in the sample.
    [2]

    (c) Find the probability that there is at most two defective lamps in the sample.
    [2]

4. The probability of obtaining heads on a biased coin is 0.69. The coin is tossed nine times.

    (a) Find the mean number of heads.
    [2]

    (b) Find the probability of obtaining **exactly** six heads.
    [2]

    (c) Find the probability of obtaining **less than** three heads.
    [2]

84 Paper 1 – Unknown Success Probability

Example

Sam likes to play a game of chance. For this game, the probability that Sam wins is p. He plays the game ten times.

(a) Write down an expression, in terms of p, for the probability that he wins exactly three games.

[2]

(b) Hence, find the possible values of p such that the probability that he wins exactly three games is 0.22.

[3]

Solution

(a) The required probability

$$=\binom{10}{3}p^3(1-p)^{10-3} \quad \text{(A1) for substitution}$$

$$=\binom{10}{3}p^3(1-p)^7 \quad \text{A1} \quad \text{N2}$$

[2]

(b) $\binom{10}{3}p^3(1-p)^7 = 0.22$ (M1) for setting equation

$\binom{10}{3}p^3(1-p)^7 - 0.22 = 0$

By considering the graph of

$y = \binom{10}{3}p^3(1-p)^7 - 0.22$,

$p = 0.216$ and $p = 0.394$ \quad A2 \quad N3

[3]

Your Practice Set – Applications and Interpretation for IBDP Mathematics

Exercise 84

1. A factory makes switches. The probability that a switch is defective is p. The factory tests a random sample of 120 switches.

    (a) Write down an expression, in terms of p, for the probability that there are exactly three defective switches in the sample.
    [2]

    (b) Hence, find the possible values of p such that the probability that there are exactly three defective switches in the sample is 0.16.
    [3]

2. Sandy goes to work five days a week. On any day, the probability that she goes on a bus is p.

    (a) Write down an expression, in terms of p, for the probability that she goes to work on a red bus on exactly four days.
    [2]

    (b) Hence, find the possible values of p such that the probability that she goes to work on a red bus on exactly four days is 0.3.
    [3]

3. The probability of obtaining "tails" when a biased coin is tossed is q. The coin is tossed ten times.

    (a) Write down an expression, in terms of q, for the probability of obtaining at least nine tails.
    [2]

    (b) Hence, find the value of q such that the probability of obtaining at least nine tails is 0.09.
    [3]

4. There are one hundred customers in a café in a particular day. The probability that a customer buys chocolate muffins is q.

    (a) Write down an expression, in terms of q, for the probability that there are at most one customer buying chocolate muffins.
    [2]

    (b) Hence, find the value of q such that the probability that there are at most one customer buying chocolate muffins is 0.03.
    [3]

85 Paper 2 – Unknown Number of Trials

Example

Ruby goes to school five days a week. When it rains, the probability that she goes to school by taxi is 0.7. When it does not rain, the probability that she goes to school by taxi is 0.1. The probability that it rains on any given day is 0.32.

(a) On a randomly selected school day, find the probability that Ruby goes to school by taxi.

[3]

(b) Given that Ruby went to school by taxi on Friday, find the probability that it was raining.

[3]

(c) In a randomly chosen school week, find the probability that Ruby goes to school by taxi on exactly two days.

[3]

(d) After n school days, the probability that Ruby goes to school by taxi at least once is greater than 0.9. Find the least value of n.

[5]

Solution

(a) The required probability
$= (0.32)(0.7) + (1 - 0.32)(0.1)$ (M1)(A1) for substitution
$= 0.292$ A1   N3

[3]

(b) The required probability
$= \dfrac{(0.32)(0.7)}{0.292}$ (M1)(A1) for substitution
$= 0.767$ A1   N3

[3]

(c) $X \sim B(5, 0.292)$ (R1) for binomial distribution
$P(X = 2)$
$= \binom{5}{2}(0.292)^2(1 - 0.292)^{5-2}$ (A1) for substitution
$= 0.303$ A1   N3

[3]

Your Practice Set – Applications and Interpretation for IBDP Mathematics

(d) The probability that Ruby does not go to school by taxi

$= (1 - 0.292)^n$ (M1) for valid approach

$= 0.708^n$

$1 - 0.708^n > 0.9$ (M1)(A1) for correct inequality

$0.1 - 0.708^n > 0$

By considering the graph of $y = 0.1 - 0.708^n$,

$n > 6.6681451$ (A1) for correct value

$\therefore n = 7$ A1   N5

[5]

Exercise 85

1. There are six working days for Joyce per week. She can either choose going back to office to work or stay at home to work. When it snows, the probability that she goes to office is 0.12. When it does not snow, the probability that she goes to office is 0.76. The probability that it snows on any given day is 0.56.

    (a) On a randomly selected working day, find the probability that Joyce goes to office.

    [3]

    (b) Given that Joyce goes to office to work on a particular day, find the probability that it was not snowing.

    [3]

    (c) In a randomly chosen working week, find the probability that Joyce stay at home to work on exactly four days.

    [3]

    (d) After n working days, the probability that Joyce stay at home to work at least once is greater than 0.84. Find the least value of n.

    [5]

2. Anson goes to school four days a week. He travels to school either by car or by bicycle. On any particular day the probability that he travels by car is 0.4.

    The probability of being late for school is 0.2 if he travels by car.

    The probability of being late for school is 0.3 if he travels by bicycle.

    (a) On a randomly selected school day, find the probability that Anson is late.

    [3]

    (b) Given that Anson is late on a particular school day, find the probability that he travels by bicycle.

    [3]

(c) In a randomly chosen school week, find the probability that Anson is late on exactly two days.

[3]

(d) After n school days, the probability that Anson is late more than once is greater than 0.75. Find the least value of n.

[5]

3. Lydia notices that on windy days, the probability she catches a fish is 0.13 while on non-windy days the probability she catches a fish is 0.59. The probability that it will be windy on a particular day is 0.45.

(a) On a randomly selected day, find the probability that Lydia catches a fish.

[3]

(b) Given that Lydia catches a fish on a particular day, find the probability that it was not windy.

[3]

(c) In a week of seven days, find the probability that Lydia catches a fish on exactly three days.

[3]

(d) After n days, the probability that Lydia catches a fish on at least two days is greater than 0.93. Find the least value of n.

[5]

4. In an experiment, a mouse is placed in a maze. There are two doors, A and B, and the mouse has to choose one of them to escape from the maze. Both doors are linked to various paths. Some paths in the maze lead to a trap and others to escape doors. The probability that the mouse chooses door A is p. If it chooses the door A, then the probability that it reaches a trap is 0.7. If it chooses the door B, then the probability that it reaches a trap is 0.52.

(a) Find the probability that the mouse reaches an escape door, giving the answer in terms of p.

[3]

(b) Given that the mouse reaches an escape door, write down the probability that it chooses the door A, giving the answer in terms of p.

[2]

Assume that $p = 0.61$.

(c) If the experiment is repeated for eight times, find the probability that the mouse reaches an escape door on exactly six times.

[3]

(d) After n experiments, the probability that the mouse reaches an escape door on more than two trials is greater than 0.99. Find the least value of n.

[5]

Chapter 21

Normal Distribution

SUMMARY POINTs

- Properties of a random variable $X \sim N(\mu, \sigma^2)$ following normal distribution:
  1. μ : Mean
  2. σ : Standard deviation
  3. The mean, the median and the mode are the same
  4. The normal curve representing the distribution is a bell-shaped curve which is symmetric about the middle vertical line
  5. $P(X < \mu) = P(X > \mu) = 0.5$
  6. The total area under the curve is 1

Solutions of Chapter 21

86 Paper 1 – Symmetric Properties of a Normal Curve

Example

The random variable X is normally distributed with a mean of 45. The following diagram shows the normal curve for X.

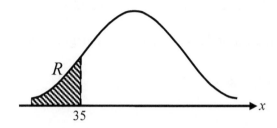

Let R be the shaded region under the curve, to the left of 35. The area of R is 0.18.

(a) Write down $P(X < 35)$.

[1]

(b) Find $P(35 < X < 45)$.

[3]

(c) Find $P(45 < X < 55)$.

[2]

Solution

(a) $P(X < 35) = 0.18$ A1   N1

[1]

(b) $P(35 < X < 45)$
$= P(X < 45) - P(X < 35)$ (M1) for valid approach
$= 0.5 - 0.18$ (A1) for substitution
$= 0.32$ A1   N3

[3]

(c) $P(45 < X < 55)$
$= P(35 < X < 45)$ (M1) for symmetric property
$= 0.32$ A1   N2

[2]

Your Practice Set – Applications and Interpretation for IBDP Mathematics

Exercise 86

1. The random variable X is normally distributed with a mean of 80. The following diagram shows the normal curve for X.

   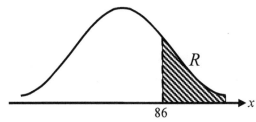

   Let R be the shaded region under the curve, to the right of 86. The area of R is 0.28.

   (a) Write down $P(X > 86)$.

   [1]

   (b) Find $P(80 < X < 86)$.

   [3]

   (c) Find $P(74 < X < 80)$.

   [2]

2. The random variable X is normally distributed with a mean of 300. The following diagram shows the normal curve for X.

   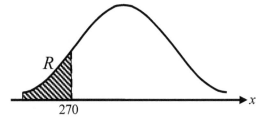

   Let R be the shaded region under the curve, to the left of 270. The area of R is 0.15.

   (a) Write down $P(X < 270)$.

   [1]

   (b) Find $P(270 < X < 300)$.

   [3]

   (c) Find $P(270 < X < 330)$.

   [2]

3. The random variable X is normally distributed with a mean of 1.5. The following diagram shows the normal curve for X.

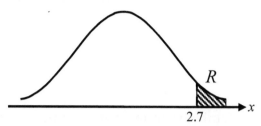

Let R be the shaded region under the curve, to the right of 2.7. The area of R is 0.07.

(a) Write down $P(X > 2.7)$.

[1]

(b) Find $P(1.5 < X < 2.7)$.

[3]

(c) Find $P(X > 0.3)$.

[2]

4. The random variable X is normally distributed with a mean of $\dfrac{6}{11}$. The following diagram shows the normal curve for X.

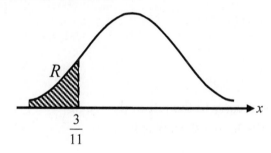

Let R be the shaded region under the curve, to the left of $\dfrac{3}{11}$. The area of R is $\dfrac{1}{6}$.

(a) Write down $P\left(X > \dfrac{3}{11}\right)$.

[2]

(b) If $P(X > d) = P\left(X < \dfrac{3}{11}\right)$, find d.

[2]

(c) Find $P\left(\dfrac{3}{11} < X < d\right)$.

[2]

87 Paper 1 – Critical Values in a Normal Distribution

Example

A random variable X is distributed normally with a mean of 100 and standard deviation of 5.

(a) On the following diagram, shade the region representing $P(X \leq 97)$.

[2]

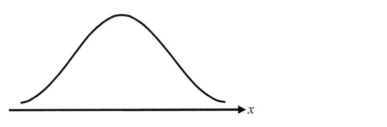

(b) Write down $P(X \leq 97)$, correct to four decimal places.

[2]

(c) Let $P(X \leq c) = 0.4$. Write down the value of c.

[2]

Solution

(a) For vertical line clearly to the left of the mean   A1
      For shading to the left of the vertical line   A1   N2

[2]

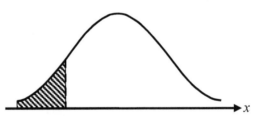

(b) $P(X \leq 97) = 0.2742530646$ (A1) for correct value
       $P(X \leq 97) = 0.2743$ A1   N2

[2]

(c) $c = 98.73326449$
       $c = 98.7$ A2   N2

[2]

Exercise 87

1. A random variable X is distributed normally with a mean of 65 and standard deviation of 2.5.

   (a) On the following diagram, shade the region representing $P(X \leq 60)$.

   [2]

   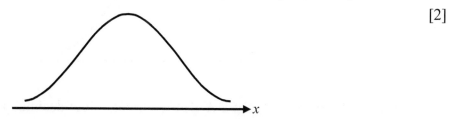

   (b) Write down $P(X \leq 60)$, correct to four decimal places.

   [2]

   (c) Let $P(X \leq c) = 0.1$. Write down the value of c.

   [2]

2. A random variable X is distributed normally with a mean of 4.41 and standard deviation of 0.67.

   (a) On the following diagram, shade the region representing $P(X \geq 4.83)$.

   [2]

   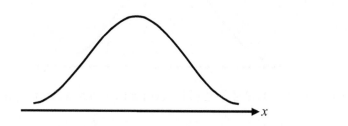

   (b) Write down $P(X \geq 4.83)$, correct to four decimal places.

   [2]

   (c) Let $P(X \geq c) = 0.3$. Write down the value of c.

   [2]

Your Practice Set – Applications and Interpretation for IBDP Mathematics

3. A random variable X is distributed normally with a mean of 30 and standard deviation of 4.

    (a) On the following diagram, shade the region representing $P(23.5 \leq X \leq 30)$.

    [2]

    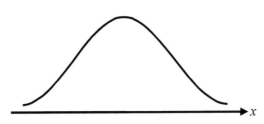

    (b) Write down $P(23.5 \leq X \leq 30)$, correct to four decimal places.

    [2]

    (c) Let $P(30 \leq X \leq c) = 0.2$. Write down the value of c.

    [2]

4. A random variable X is distributed normally with a mean of 162 and standard deviation of 8.

    (a) On the following diagram, shade the region representing $P(162 \leq X \leq 171)$.

    [2]

    (b) Write down $P(162 \leq X \leq 171)$, correct to four decimal places.

    [2]

    (c) Let $P(c \leq X \leq 162) = 0.19$. Write down the value of c.

    [2]

88 Paper 1 – Quartiles of a Normal Distribution

Example

A random variable X is normally distributed with $\mu = 70$ and $\sigma = 6$.

(a) Find the lower quartile of X.

[2]

(b) Find the upper quartile of X.

[2]

(c) Hence, find the interquartile range of X.

[2]

Solution

(a) $P(X < Q_1) = 0.25$ (M1) for valid approach

$Q_1 = 65.9530615$

Thus, the lower quartile is 66.0. A1 N2

[2]

(b) $P(X < Q_3) = 0.75$ (M1) for valid approach

$Q_3 = 74.0469385$

Thus, the upper quartile is 74.0. A1 N2

[2]

(c) The interquartile range of X

$= Q_3 - Q_1$ (A1) for correct formula

$= 74.0469385 - 65.9530615$

$= 8.093876997$

$= 8.09$ A1 N2

[2]

Your Practice Set – Applications and Interpretation for IBDP Mathematics

Exercise 88

1. A random variable X is normally distributed with $\mu = 240$ and $\sigma = 15$.

    (a) Find the lower quartile of X.
    [2]

    (b) Find the upper quartile of X.
    [2]

    (c) Hence, find the interquartile range of X.
    [2]

2. A random variable X is normally distributed with $\mu = 160$ and $\sigma = 8$.

    (a) Find the value of the 30th percentile of X.
    [2]

    (b) Find the value of the 70th percentile of X.
    [2]

    (c) Hence, find the difference between the 70th percentile and the 30th percentile.
    [2]

3. A random variable X is normally distributed with $\mu = 88$ and $\sigma = 4$. Find the difference between the 90th percentile and the 10th percentile.
    [6]

4. A random variable X is normally distributed with $\mu = 50$ and $\sigma = 3$. The values s and t, where $s < t$, are the data values that divide the whole distribution into 3 equal parts. Find $t - s$.
    [6]

89 Paper 2 – Real Life Problems

Example

The heights of the trees in a park are normally distributed with a mean of 60 metres and a standard deviation of 6 metres. Trees are classified as short trees if they are shorter than 52.5 metres.

(a) A tree is selected at random from the park.

 (i) Find the probability that this tree is a short tree.

 (ii) Given that this tree is a short tree, find the probability that it is shorter than 50 metres.

[6]

(b) Two trees are selected at random. Find the probability that they are both short trees.

[2]

(c) 80 trees are selected at random.

 (i) Find the expected number of these trees that are short trees.

 (ii) Find the probability that less than 11 of these trees are short trees.

[6]

Your Practice Set – Applications and Interpretation for IBDP Mathematics

Solution

(a) (i) Let H : Height of a tree in the park
The required probability
$= P(H < 52.5)$ (M1) for valid approach
$= 0.105649839$
$= 0.106$ A1 N2

(ii) $P(H < 50 \mid H < 52.5)$ (R1) for correct probability

$= \dfrac{P(H < 50 \cap H < 52.5)}{P(H < 52.5)}$ (A1) for correct formula

$= \dfrac{P(H < 50)}{P(H < 52.5)}$

$= \dfrac{0.0477903304}{0.105649839}$ (A1) for correct values

$= 0.4523464571$
$= 0.452$ A1 N4

[6]

(b) The required probability
$= P(H < 52.5) \times P(H < 52.5)$ (M1) for valid approach
$= 0.105649839 \times 0.105649839$
$= 0.0111618885$
$= 0.0112$ A1 N2

[2]

(c) (i) The required expected number
$= (80)(0.105649839)$ (A1) for correct formula
$= 8.45198712$
$= 8.45$ A1 N2

(ii) Let X : Number of short trees in the selected sample
$X \sim B(80, 0.105649839)$ (R1) for binomial distribution
The required probability
$= P(X < 11)$ (M1) for valid approach
$= P(X \leq 10)$ (A1) for correct value
$= 0.7784376955$
$= 0.778$ A1 N4

[6]

Exercise 89

1. The weights of fish in a lake are normally distributed with a mean of 740 g and standard deviation 46 g. Fishes are classified as big fishes if they are heavier than 850 g.

    (a) A fish is randomly selected from the lake.

        (i) Find the probability that this fish is big.

        (ii) Given that this fish is big, find the probability that it is heavier than 900 g.

    [6]

    (b) Two fishes are randomly selected. Find the probability that they are both big.

    [2]

    (c) 100 fishes are randomly selected.

        (i) Find the expected number of these fishes that are big.

        (ii) Find the probability that more than 2 of these fishes are big.

    [6]

2. A company produces containers of milk soda. The volume of milk soda in the containers is normally distributed with a mean of 350 ml and standard deviation of 6 ml.

    A container which contains less than 335 ml of milk soda is unsatisfied.

    (a) A container is selected at random.

        (i) Find the probability that this container is unsatisfied.

        (ii) Given that this container is unsatisfied, find the probability that it contains more than 330 ml of milk soda.

    [6]

    (b) Two containers are selected at random. Find the probability that only one of them is unsatisfied.

    [3]

    (c) 60 containers are selected at random.

        (i) Find the expected number of these containers that are unsatisfied.

        (ii) Find the probability that less than 3 of these containers are unsatisfied.

    [6]

3. A machine manufactures a large number of nails. The length, L mm, of a nail is normally distributed, where $L \sim N(70, 3^2)$. There are 15% of the nails which are shorter than t mm, and they are classified as short nails.

    (a) A nail is randomly selected.

    (i) Find t.

    (ii) Given that this nail is short, find the probability that it is shorter than 65 mm.
    [6]

    (b) Two nails are randomly selected. Find the probability that only one of them is short.
    [3]

    (c) 25 nails are randomly selected. Let X be the number of short nails in this sample.

    (i) Find the variance of X.

    (ii) Find the probability that at least 4 of these nails are short.
    [6]

4. The masses of watermelons grown on a farm are normally distributed with a mean of 9 kg and standard deviation 0.4 kg. A watermelon is large if its mass is greater than t kg. Ten percent of the watermelons are classified as large.

    (a) A watermelon is selected at random.

    (i) Find t.

    (ii) Given that this watermelon is large, find the probability that it is less than 9.8 kg.
    [6]

    (b) Three watermelons are selected at random. Find the probability that they are all large.
    [2]

    (c) 52 watermelons are selected at random. Let X be the number of large watermelons in this sample.

    (i) Find the variance of X.

    (ii) Find the probability that at least 13 and at most 26 of these watermelons are large.
    [6]

Chapter 22

Bivariate Analysis

SUMMARY POINTs

- Correlations:

Positive	Strong	$0.75 < r < 1$
	Moderate	$0.5 < r < 0.75$
	Weak	$0 < r < 0.5$
No		$r = 0$
Negative	Weak	$-0.5 < r < 0$
	Moderate	$-0.75 < r < -0.5$
	Strong	$-1 < r < -0.75$

where r is the correlation coefficient

- Correlation Coefficient for ranked data:
  r_s : Spearman's Rank Correlation Coefficient

Solutions of Chapter 22

Paper 1 – Scatter Diagrams

Example

The following scatter diagram shows the marks scored by six students on two different assessments, Assessment 1 (x) and Assessment 2 (y):

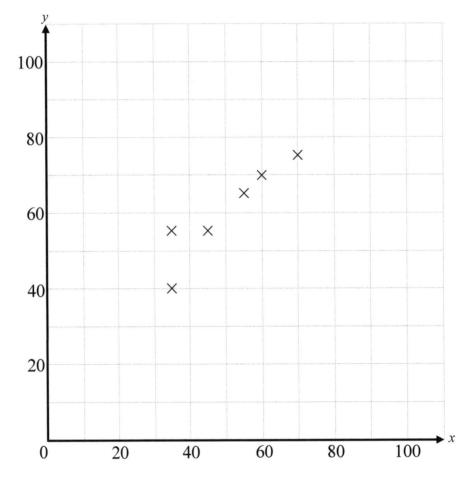

(a) Plot and label the mean point $M(50, 60)$ on the scatter diagram.

[2]

(b) Draw a suitable line of best fit on the scatter diagram.

[2]

(c) Using the line of best fit to estimate the marks scored by a student in the Assessment 2 when the corresponding marks scored in the Assessment 1 is 0, giving the answer correct to the nearest integer.

[2]

Solution

(a) For correct position   A1
     For labelling   A1   N2

[2]

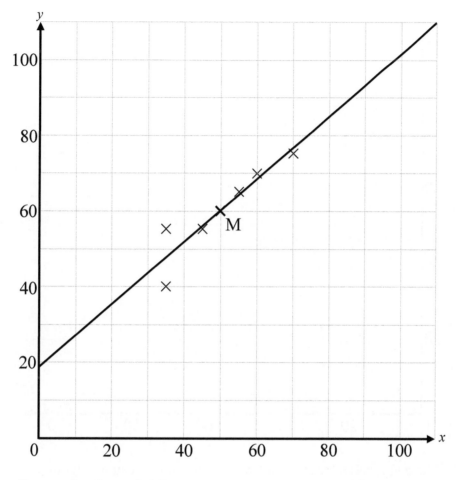

(b) For passing through M   A1
     For extending the line to the y-axis   A1   N2

[2]

(c) 19   A2   N2

[2]

1. The following scatter diagram shows the distributions of the values of x and y.

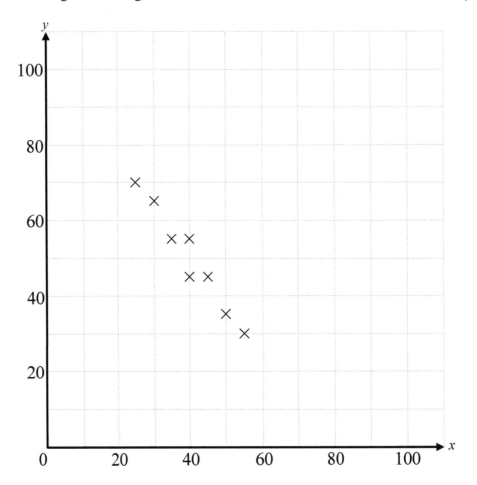

(a) Plot and label the mean point M(40, 50) on the scatter diagram.

[2]

(b) Draw a suitable line of best fit on the scatter diagram.

[2]

(c) Describe the correlation between x and y.

[2]

(d) Using the line of best fit to estimate the value of y when $x = 20$.

[2]

2. The following table below shows the distance d in kilometres and the cost c in pounds of each of six taxi journeys.

Distance (d km)	50	55	65	75	85	90
Cost (c pounds)	55	70	60	80	70	85

The above data is visualized by the scatter diagram below, with one point missing.

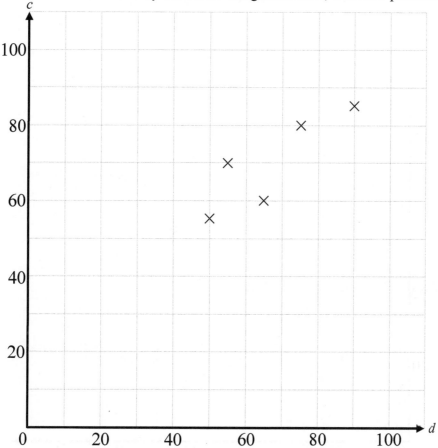

(a) Plot and label the missing point P on the scatter diagram.

[2]

(b) Write down the coordinates of the mean point M.

[2]

(c) Plot and label the mean point M on the scatter diagram.

[1]

(d) Draw a suitable line of best fit on the scatter diagram.

[2]

3. The following table shows the Art exam score x and the Music exam score y for six students.

Art exam score (x)	95	90	80	80	65	70
Music exam score (y)	95	85	70	80	55	65

The above data is visualized by the scatter diagram below:

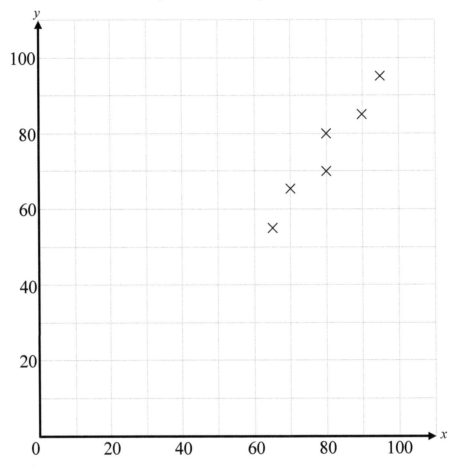

(a) Plot and label the mean point M(80, 75) on the scatter diagram.

[2]

(b) Draw a suitable line of best fit on the scatter diagram.

[2]

This line of best fit can be used to make reliable estimates of the music exam score for a known art exam score of a student, when this known art exam score lies between p and q, where $p < q$.

(c) Write down the values of p and q.

[2]

4. The following table below shows the age x of a person and the score y on a memory test of seven patients.

Age (x)	20	20	25	30	35	40	40
Score (y)	70	60	55	45	35	30	20

The above data is visualized by the scatter diagram below, where M(30, 45) represents the mean point:

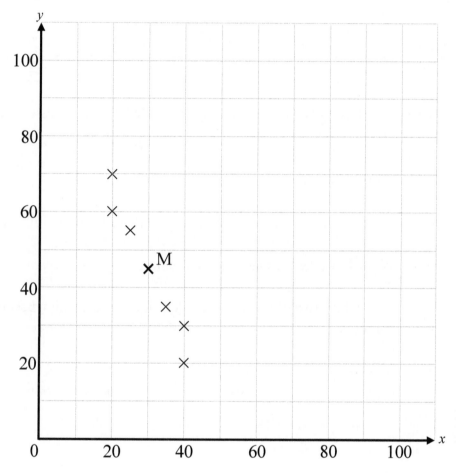

(a) Draw a suitable line of best fit on the scatter diagram.

[2]

(b) Using the line of best fit to estimate the memory test score of a patient when the patient is 45 years old, giving the answer correct to the nearest integer.

[2]

This line of best fit would make unreliable estimates of the memory test score for patients under the age of p or above the age of q, where $p < q$.

(c) Write down the values of p and q.

[2]

Your Practice Set – Applications and Interpretation for IBDP Mathematics

Paper 1 – Describe the Correlations

Example

The following table shows the average weights (y kg) for given heights (x cm) in a group of people.

Heights (x cm)	156	160	164	168	172	176
Weights (y kg)	52	58	59	65	67	72

(a) The relationship between the variables is modelled by the regression equation $y = ax + b$.

    (i) Write down the value of a and of b.

    (ii) Hence, estimate the weight of a person whose height is 170 cm.

[4]

(b) (i) Write down the correlation coefficient.

    (ii) State which **two** of the following describe the correlation between the variables.

[3]

    positive   strong   zero
    negative   weak   no correlation

Solution

(a) (i) $a = 0.95$ A1   N1
         $b = -95.53333333$
         $b = -95.5$ A1   N1

    (ii) The estimated weight
         $= 0.95(170) - 95.53333333$ (A1) for substitution
         $= 65.96666667$
         $= 66.0$ kg   A1   N2

[4]

(b) (i) $r = 0.9880809318$
         $r = 0.988$ A1   N1

    (ii) Strong, Positive   A2   N2

[3]

Exercise 91

1. The following table shows the final exam scores for given midterm test scores in a group of students.

Midterm test scores (x)	80	83	86	89	92
Final exam scores (y)	70	71	67	63	77

   (a) The relationship between the variables is modelled by the regression equation $y = ax + b$.

       (i) Write down the value of a and of b.

       (ii) Hence, estimate the final exam score of a student whose midterm test score is 85.

   [4]

   (b) (i) Write down the correlation coefficient.

       (ii) State which **two** of the following describe the correlation between the variables.

   [3]

       positive   strong   zero
       negative   weak   no correlation

2. The following table shows the temperature y degrees Celsius of a liquid, x hours after the start of an experiment.

x	0	2	4	6	8	10
y	90	92	88	103	70	71

   (a) The relationship between the variables is modelled by the regression equation $y = ax + b$.

       (i) Write down the value of a and of b.

       (ii) Hence, estimate the temperature of the liquid 9 hours after the start of the experiment.

   [4]

(b) (i) Write down the correlation coefficient.

(ii) State which **two** of the following describe the correlation between the variables.

[3]

| positive | strong | zero |
| negative | weak | moderate |

3. The following table shows the Mathematics test scores (x) and the Physics test scores (y) of 7 students.

Mathematics test scores (x)	14	19	16	20	30	17	11
Physics test scores (y)	16	19	18	23	30	16	19

(a) The relationship between the variables is modelled by the regression equation $y = ax + b$.

(i) Write down the value of a and of b.

(ii) Hence, estimate the Physics test score of a student whose Mathematics test score is 25.

[4]

A teacher believes that the relationship between the Mathematics test scores (x) and the Chemistry test scores (w) of the students can be modelled by a linear regression equation.

(b) The teacher describes the correlation as **very strong and negative**. Circle the value below which best represents the correlation coefficient.

[1]

0.989   0.178   0   −0.178   −0.989

(c) The teacher's model is $w = -0.18x + 25.6$. Use his model to predict the difference in Chemistry test scores of two students if the difference of their Mathematics test scores is 3.

[2]

4. The following table shows the average number of hours per day spent watching television by the mother and the youngest child of six families.

Hours per day spent watching television by a mother (x)	1.8	2.1	2.2	2.4	2.7	3.8
Hours per day spent watching television by her child (y)	5.5	5.4	4.9	4.9	4.6	4.0

The relationship can be modelled by the regression line with equation $y = ax + b$.

(a) (i) Find the correlation coefficient r.

(ii) Write down the value of a and of b.

[3]

A social worker believes that the relationship between the average number of hours per day spent watching television by the mother (x) of a family and average number of hours per day spent watching television by her husband (w) can be modelled by a linear regression equation.

(b) The social worker describes the correlation as **weak and positive**. Circle the value below which best represents the correlation coefficient.

[1]

0.989   0.178   0   −0.178   −0.989

(c) The social worker's model is $w = 0.53x$. Use her model to predict the average number of hours per day spent watching television by the father if his wife watches television for an average of 2.7 hours per day.

[2]

92 Paper 2 – Coefficients of a Regression Line

Example

The following table shows the amount of fuel (y litres) used by a vehicle to travel certain distances (x km).

Travel distance (x km)	5	8	11	16
Amount of fuel (y litres)	0.5	0.7	1.2	1.3

This data can be modelled by the regression line with equation $y = ax + b$.

(a) (i) Write down the value of a and of b.

    (ii) Explain what the gradient a represents.

[3]

(b) Use the model to estimate the amount of fuel the vehicle would use if its travelling distance is 13 km.

[2]

Solution

(a) (i) $a = 0.0772727273$
           $a = 0.0773$ A1   N1
           $b = 0.1522727273$
           $b = 0.152$ A1   N1

    (ii) a represents the average amount of fuel required to travel 1 km   A1   N1

[3]

(b) The estimated amount of fuel
$= 0.0772727273(13) + 0.1522727273$ (A1) for substitution
$= 1.156818182$
$= 1.16$ litres   A1   N2

[2]

Exercise 92

1. The following table shows the public exam score x and university entrance mark y for five students.

Public exam score (x)	172	178	185	190	191
University entrance mark (y)	21.2	22.4	22.5	24.5	24.1

   This data can be modelled by the regression line with equation $y = ax + b$.

   (a) (i) Write down the value of a and of b.

   (ii) Explain what the gradient a represents.
   [3]

   (b) Sarah scored a total of 180 in her public exam. Use the model to estimate her university entrance mark.
   [2]

2. The following table shows the sales, y millions of dollars, of an online shop, x years after 2011.

Number of years after 2011 (x years)	1	2	3	4	5	6
Sales (y millions of dollars)	6.8	7.7	9.9	13.7	21.2	21.9

   The relationship between the variables modelled by the regression line with equation $y = ax + b$.

   (a) (i) Find the value of a and of b.

   (ii) Explain what the intercept b represents.
   [3]

   (b) Hence estimate the sales in millions of dollars thirty months after 2011.
   [2]

Your Practice Set – Applications and Interpretation for IBDP Mathematics

3. The maximum temperature t, in degrees Celsius, in a monument on four randomly selected days is shown in the following table. The table also shows the number of visitors, n, to the monument on each of those four days.

Maximum temperature (t degrees Celsius)	2	5	7	1
Number of visitors (n)	50	48	63	15

The relationship between the variables can be modelled by the regression equation $n = at + b$.

(a) (i) Find the value of a and of b.

    (ii) Explain what a and b represent.

[4]

(b) Use the regression equation to estimate the number of visitors on a day when the maximum temperature is 4°C.

[2]

4. Six metal ingots were chosen at random and measurements were made of their breaking strength s and their hardness h. The results are shown in the table below.

Breaking strength (s tonnes per cm)	5	7	7.4	6.8	5.5	6.9
Hardness (h)	51	71	83	70	75	60

The relationship between the variables can be modelled by the regression equation $h = as + b$.

(a) (i) Find the value of a and of b.

    (ii) Explain what a and b represent.

[4]

(b) Use the regression equation to estimate the hardness of a metal ingot when its breaking strength is six tonnes per cm.

[2]

93 Paper 1 – Spearman's Rank Correlation Coefficient

Example

The table below shows the ranking of eight universities suggested by two educational experts, where 1 represents the highest ranking and 8 represents the lowest ranking:

University	A	B	C	D	E	F	G	H
Expert 1	2	3	1	7	8	4	4	6
Expert 2	1	2	2	8	7	4	6	5

(a) Write down

    (i) the average equal ranks of the University F and the University G in Expert 1's ranking;

    (ii) the average equal ranks of the University B and the University C in Expert 2's ranking.

[2]

(b) Find the Spearman's rank correlation coefficient for the above data set.

[2]

(c) Comment on the conclusion of the result in (b).

[1]

Solution

(a) (i)   4.5   A1   N1

    (ii)   2.5   A1   N1

[2]

(b) $r_s = 0.8915662651$ (A1) for correct value

       $r_s = 0.892$ A1   N2

[2]

(c) There is a strong agreement between the two experts.   A1   N1

[1]

Your Practice Set – Applications and Interpretation for IBDP Mathematics

Exercise 93

1. Ravi and Yannick are trying to design a quiz on Statistics. They designed six questions and they are going to comment on the relative difficulties of the questions. The table below shows the ranking of six questions suggested by them, where 1 represents the most difficult question and 6 represents the easiest question:

Question	A	B	C	D	E	F
Ravi	4	5	3	2	5	1
Yannick	4	6	2	4	1	3

   (a) Write down

   (i) the average equal ranks of the Question B and the Question E in Ravi's ranking;

   (ii) the average equal ranks of the Question A and the Question D in Yannick's ranking.
   [2]

   (b) Find the Spearman's rank correlation coefficient for the above data set.
   [2]

   (c) Comment on the conclusion of the result in (b).
   [1]

2. An equestrian event is held in the Summer Olympics in 2012. The table below shows the ranking of eight horse riders suggested by two judges, where 1 represents the highest ranking and 8 represents the lowest ranking:

Horse rider	A	B	C	D	E	F	G	H
Judge 1	1	5	2	7	7	4	6	3
Judge 2	1	4	4	8	7	2	6	3

   (a) Find the Spearman's rank correlation coefficient for the above data set.
   [4]

   (b) Comment on the conclusion of the result in (b).
   [1]

3. The number of goals of eight teams in a football league and their corresponding positions are summarized in the following table:

Team	A	B	C	D	E	F	G	H
Number of goals	68	62	32	49	15	45	46	28
Position in the league	3	1	6	2	8	4	5	7

In the table below, the number of goals for each team is going to be ranked, such that 1 represents the team with the most number of goals and 8 represents the team with the least number of goals.

Team	A	B	C	D	E	F	G	H
Number of goals' rankings	1	b	6	d	8	f	g	7
Position in the league	3	1	6	2	8	4	5	7

(a) Write down the values of b, d, f and g.

[2]

(b) Find the Spearman's rank correlation coefficient for the second table.

[2]

(c) Comment on the conclusion of the result in (b).

[1]

4. The following table shows the rankings of the six observations for two variables X and Y, where 1 represents the highest rank and 6 represents the lowest rank:

Observation	First	Second	Third	Fourth	Fifth	Sixth
X	2	4	3	5	6	1
Y	1	2	3	4	5	6

(a) Find the Spearman's rank correlation coefficient for the above data set.

[2]

The sixth observation of both X and Y are said to be the outlier of the data set and are deleted. The following table shows the updated rankings:

Observation	First	Second	Third	Fourth	Fifth
X	a	b	c	d	e
Y	1	2	3	4	5

(b) (i) Write down the values of a and e.

(ii) Find the value of the R_s, the Spearman's rank correlation coefficient for the updated data set.

(iii) Comment on the change in the values of the Spearman's rank correlation coefficient.

[4]

Your Practice Set – Applications and Interpretation for IBDP Mathematics

 Paper 1 – Two Different Correlation Coefficients

Example

The following table shows the values of the seven observations for two variables X and Y:

Observation	First	Second	Third	Fourth	Fifth	Sixth	Seventh
X	2	4	6	8	10	12	14
Y	1	2	4	8	28	20	62

(a) Explain why it is not appropriate to use the product moment correlation coefficient for the above table.

[1]

In order to calculate the Spearman's rank correlation coefficient, the rankings of the seven observations for two variables X and Y are assigned, where 1 represents the highest rank and 7 represents the lowest rank.

(b) Find the Spearman's rank correlation coefficient for the above data set.

[2]

(c) Comment on the conclusion of the result in (b).

[1]

Solution

(a) The values of the seven observations are not linear. A1   N1

[1]

(b) $r_s = 0.9642857143$ (A1) for correct value

$r_s = 0.964$ A1   N2

[2]

(c) There is a strong agreement between the two rankings.   A1   N1

[1]

Exercise 94

1. 100 customers completed the survey about their feedbacks on five different types of ice cream of different selling prices. The following table shows the overall rankings of five different types of ice cream based on the feedbacks and the corresponding selling prices, where 1 represents the highest rank and 5 represents the lowest rank:

Brand	A	B	C	D	E
Ranking	5	2	4	1	3
Selling price	$10	$12	$7	$20	$13

   (a) Explain why it is not appropriate to use the product moment correlation coefficient for the above table.

   [1]

   (b) Find the Spearman's rank correlation coefficient for the above data set.

   [2]

   (c) Comment on the conclusion of the result in (b).

   [1]

2. The following table shows the scores of a Geography test and a Chemistry test of eight students, both out of 100:

Student	A	B	C	D	E	F	G	H
Geography	90	71	60	55	53	50	49	54
Chemistry	20	30	35	40	50	45	65	70

   (a) Explain why it is not appropriate to use the product moment correlation coefficient for the above table.

   [1]

   (b) Find the Spearman's rank correlation coefficient for the above data set.

   [2]

   (c) Comment on the conclusion of the result in (b).

   [1]

3. The number of winning matches of eight teams in a football league and their corresponding positions are summarized in the following table, where x represents the number of winning matches of the Team A:

Team	A	B	C	D	E	F	G	H
Number of winning matches	x	7	7	17	6	20	12	14
Position in the league	7	5	6	2	8	1	3	4

(a) Explain why it is not appropriate to use the product moment correlation coefficient for the above table.

[1]

In the table below, the number of winning matches for each team is going to be ranked, such that 1 represents the team with the most number of winning matches and 8 represents the team with the least number of winning matches.

Team	A	B	C	D	E	F	G	H
Number of winning matches' rankings	8	p	p	2	q	1	4	3
Position in the league	7	5	6	2	8	1	3	4

(b) Write down the values of p and q.

[2]

(c) Find the Spearman's rank correlation coefficient for the second table.

[2]

The Spearman's rank correlation coefficient found in (c) is unchanged for a range of values of x.

(d) Write down all possible values of x.

[1]

4. Seven divers participated in a diving competition. There are two judges in total: A trainee judge and a professional judge. The following table shows the scores of the seven divers awarded by the two judges:

Diver	A	B	C	D	E	F	G
Trainee judge	8.6	8.9	8.7	8.5	8.2	8.4	8.1
Professional judge	8.5	8.8	8.7	8.4	8.3	8.0	8.2

(a) Find the product moment correlation coefficient for the above data set.

[2]

In the table below, the scores of the seven divers awarded by the two judges are going to be ranked, such that 1 represents the highest score and 8 represents the lowest score.

Diver	A	B	C	D	E	F	G
Trainee judge	3	1	2	4	6	5	7
Professional judge	a	b	c	d	e	f	g

(b) (i) Express b in terms of a.

(ii) Express g in terms of a.

[2]

(c) Find the Spearman's rank correlation coefficient for the second table.

[2]

Later, the trainee judge adjusts the score of the diver B such that the score is increased.

(d) Explain why the Spearman's rank correlation coefficient in (c) is unchanged.

[1]

Your Practice Set – Applications and Interpretation for IBDP Mathematics

Paper 2 – Miscellaneous Problems

Example

The insurance cost of a van depends partly on the distance it has travelled. The following table shows the distance and the insurance cost for six vans on 1 July 2019.

Travelling distance (x km)	10000	12500	17500	20000	27500	35000
Insurance cost ($ y)	9000	8800	8700	8600	8400	7800

The relationship between x and y can be modelled by the regression equation $y = ax + b$.

(a) (i) Find the correlation coefficient.

   (ii) Write down the value of a and of b.

[4]

On 1 July 2019, Jae Lim buys a van which has travelled 25000 km.

(b) Use the regression equation to estimate the insurance cost of Jae Lim's van, giving your answer to the nearest 100 dollars.

[3]

The insurance cost of a van decreased by 3% each year.

(c) Calculate the insurance cost of Jae Lim's van after 8 years.

[4]

Jae Lim will stop using his van when its insurance cost reaches 5000 dollars.

(d) Find the year when Jae Lim stops using his van.

[4]

Solution

(a) (i) $r = -0.9717142437$ (A1) for correct value
$r = -0.972$ A1   N2

(ii) $a = -0.0432$ A1   N1
$b = 9432$ A1   N1

[4]

(b) The estimated insurance cost
$= -0.0432(25000) + 9432$ (A1) for substitution
$= 8352$ (A1) for correct value
$= \$8400$ A1   N3

[3]

(c) The insurance cost
$= 8352 \times (1 - 3\%)^8$ (M1)(A1) for correct formula
$= 8352 \times 0.97^8$ (A1) for simplification
$= 6545.824538$
$= \$6550$ A1   N4

[4]

(d) $8352 \times (1 - 3\%)^t = 5000$ (M1) for setting equation
$8352 \times 0.97^t - 5000 = 0$ (A1) for simplification
By considering the graph of
$y = 8352 \times 0.97^t - 5000$, $t = 16.84427$.   (A1) for correct value
Thus, the year is 2036.   A1   N4

[4]

Exercise 95

1. Yuto is a beekeeper who collected data about monthly honey production in his bee hives. The data for five of his hives on 1 January 2018 is shown in the following table.

Number of bees (x)	490	620	750	815	830
Monthly honey production (y kg)	1.3	1.5	1.7	2.2	2.8

The relationship between the variables is modelled by the regression line with equation $y = ax + b$.

(a) (i) Find the correlation coefficient.

(ii) Write down the value of a and of b.

[4]

(b) Use the regression line to estimate the monthly honey production from a hive that has 700 bees, giving your answer to the nearest 0.1 kg.

[3]

Starting from 1 January 2018, the monthly honey production from the hive in (b) increased by 2% each month.

(c) Calculate the monthly honey production from the hive after 1 year.

[4]

Yuto will sell the hive to another farm when its monthly honey production reaches 3 kg.

(d) Find the year when Yuto sells the hive.

[4]

2. A farm food supplier monitors the number of chickens kept x against the monthly consumption of chicken food (y kg). The data for seven small farms on 1 July 2018 is shown in the following table.

Number of chickens (x)	10	12	14	17	19	21	25
Monthly consumption (y kg)	32	40	38	50	58	58	71

The relationship between the variables is modelled by the regression line with equation $y = ax + b$.

(a) (i) Find the correlation coefficient.

(ii) Write down the exact value of a and of b.

[4]

(b) Use the regression line to estimate the monthly consumption of chicken food when there are 24 chickens in a farm, giving your answer to the nearest kg.

[3]

Starting from 1 July 2018, the monthly consumption of chicken food from the farm in (b) increased by 5% each month.

(c) Calculate the monthly consumption of chicken food from the farm after half a year.

[4]

The supplier will enlarge the size of the farm when its monthly honey production reaches 100 kg.

(d) Find the year and the month when the supplier enlarges the size of the farm.

[4]

3. An environmental group records the numbers of wolves w and foxes f in a wildlife reserve after t years, starting on 1 January 1981 as shown in the following table.

Number of years (t)	4	8	13	18
Number of wolves (w)	257	278	382	441

The relationship between the variables is modelled by the regression line with equation $w = at + b$.

(a) (i) Find the correlation coefficient.

(ii) Write down the exact value of a and of b.

[4]

(b) Use the regression line to estimate the number of wolves on 1 January 1992, giving your answer to the nearest integer.

[3]

Let f be the number of foxes in the reserve after t years, starting on 1 January 1981. The number of foxes can be modelled by the equation $f = 50(e^{0.01kt} + 2)$, where k is a constant.

(c) After ten years, there were 930 foxes in the reserve. Find k.

[3]

(d) During which year were the number of wolves the same as the number of foxes?

[4]

4. A health clinic counted the number of breaths per minute u and the number of pulse beats per minute v for a patient t minutes after his 10 km race, as shown in the following table.

Number of minutes (t)	1	3	5	6	7
Number of breaths per minute (u)	58	55	53	50	52

The relationship between the variables is modelled by the regression line with equation $u = at + b$.

(a) (i) Find the correlation coefficient.

(ii) Write down the exact value of a and of b.

[4]

(b) Use the regression line to estimate the number of breaths per minute 12 minutes after the race, giving your answer to the nearest integer.

[3]

Your Practice Set – Applications and Interpretation for IBDP Mathematics

The number of pulse beats per minute v for the patient t minutes after his 10 km race can be modelled by the equation $v = \dfrac{10}{e^{kt}} + 70$, where k is a constant.

(c) After eight minutes, the number of pulse beats per minute is 75. Find k.

[3]

(d) After how many minutes do the number of pulse beats per minute 1.5 times the number of breaths per minute?

[4]

Chapter 23

Statistical Tests

SUMMARY POINTs

✓ χ^2 test for independence, for a contingency table with r rows and c columns:

$n = rc$: Total number of data

O_i ($i = 1, 2, \ldots, n$): Observed frequencies

E_i ($i = 1, 2, \ldots, n$): Expected frequencies

$v = (r-1)(c-1)$: Degree of freedom

$\chi^2_{calc} = \sum_{i=1}^{n} \frac{(O_i - E_i)^2}{E_i}$: χ^2 test statistic

C: Critical value in the hypothesis test

H_0: Null hypothesis

H_1: Alternative hypothesis

✓ Decision rule for χ^2 test for independence:

H_0: Two variables are independent

H_1: Two variables are not independent

H_0 is rejected if $\chi^2_{calc} > C$ or the p-value is less than the significance level

H_0 is not rejected if $\chi^2_{calc} < C$ or the p-value is greater than the significance level

Your Practice Set – Applications and Interpretation for IBDP Mathematics

SUMMARY POINTs

✓ χ^2 goodness of fit test, for a contingency table with 1 row and c columns:

$v = c - 1$: Degree of freedom

H_0: The data follows an assigned distribution

H_1: The data does not follow an assigned distribution

H_0 is rejected if $\chi^2_{calc} > C$ or the p-value is less than the significance level

H_0 is not rejected if $\chi^2_{calc} < C$ or the p-value is greater than the significance level

✓ t test:

t: t test statistic

μ_1, μ_2: The population means of two groups of data

H_0: $\mu_1 = \mu_2$

H_1: $\mu_1 > \mu_2$, $\mu_1 < \mu_2$ (for 1-tailed test), $\mu_1 \neq \mu_2$ (for 2-tailed test)

H_0 is rejected if the p-value is less than the significance level

H_0 is not rejected if the p-value is greater than the significance level

 Solutions of Chapter 23

96 Paper 1 – Goodness of Fit Test

Example

In an experiment, a fair six-sided die is tossed for 240 times. It is expected that the outcomes are evenly distributed.

The following table shows the frequencies of the outcomes:

Outcome	1	2	3	4	5	6
Frequency	36	58	48	44	30	24

A χ^2 goodness of fit test is conducted at a 5% significance level.

(a) Write down the null hypothesis of the test.

[1]

(b) Write down the value of the expected frequency of each outcome.

[1]

(c) Write down the degree of freedom of the test.

[1]

(d) Find the value of χ^2_{calc}, the test statistic.

[2]

The critical value is given by 11.070.

(e) State the conclusion of the test with a reason.

[2]

Solution

(a) H_0: The outcomes are evenly distributed.   A1   N1

[1]

(b)   40   A1   N1

[1]

(c)   5   A1   N1

[1]

(d)   19.4   A2   N2

[2]

(e)   The null hypothesis is rejected.   A1
       As $\chi^2_{calc} > 11.070$.   A1   N2

[2]

Your Practice Set – Applications and Interpretation for IBDP Mathematics

Exercise 96

1. Three different fair coins are tossed for 80 times. The following table shows the expected frequencies and the observed frequencies of the outcomes:

Number of heads	0	1	2	3
Expected frequency	10	30	30	p
Observed frequency	15	27	23	15

   A χ^2 goodness of fit test is conducted at a 5% significance level.

   (a) Write down the null hypothesis of the test.
   [1]

   (b) Write down the value of p.
   [1]

   (c) Write down the degree of freedom of the test.
   [1]

   (d) Find the value of χ^2_{calc}, the test statistic.
   [2]

   The critical value is given by 7.815.

   (e) State the conclusion of the test with a reason.
   [2]

2. In a contact list of 100 phone numbers, the last digits are recorded. It is expected that the outcomes are evenly distributed.

   The following table shows the frequencies of the outcomes:

Last digit	0	1	2	3	4	5	6	7	8	9
Frequency	10	5	17	5	6	9	13	8	15	12

   A χ^2 goodness of fit test is conducted at a 5% significance level.

   (a) Write down the null hypothesis of the test.
   [1]

   (b) Write down the value of the expected frequency of each outcome.
   [1]

   (c) Write down the degree of freedom of the test.
   [1]

   (d) Find the value of χ^2_{calc}, the test statistic.
   [2]

The critical value is given by 16.919.

(e) State the conclusion of the test with a reason.

[2]

3. In January 2020, the number of emails received on each day is recorded. The following table shows the expected frequencies and the observed frequencies of the outcomes:

Number of emails	0	1	2	3	4	5
Expected frequency	a	8	8	6	3	2
Observed frequency	2	6	6	6	6	b

A χ^2 goodness of fit test is conducted at a 5% significance level.

(a) Write down the null hypothesis of the test.

[1]

(b) Write down the values of a and b.

[2]

(c) Write down the degree of freedom of the test.

[1]

(d) Find the p-value.

[2]

(e) State the conclusion of the test with a reason.

[2]

4. In an experiment, a biased seven-sided die is tossed for 120 times. The following table shows the expected frequencies and the observed frequencies of the outcomes:

Outcome	1	2	3	4	5	6	7
Expected frequency	15	15	15	e	15	15	15
Observed frequency	16	19	8	28	9	32	8

A χ^2 goodness of fit test is conducted at a 1% significance level.

(a) Write down the null hypothesis of the test.

[1]

(b) Write down the value of e.

[1]

(c) Write down the degree of freedom of the test.

[1]

(d) Find the p-value.

[2]

(e) State the conclusion of the test with a reason.

[2]

Your Practice Set – Applications and Interpretation for IBDP Mathematics

Paper 1 – Tests for Independence

Example

Mateo is a Science teacher who wishes to investigate the relationship between the Physics grades and Chemistry grades of all 160 Science students in the school.

The following table shows the distribution of the grades (A, B and C) for the students:

		Chemistry		
		A	B	C
Physics	A	13	22	10
	B	25	30	11
	C	27	12	10

A χ^2 test for independence is conducted at a 5% significance level.

(a) (i) Write down the null hypothesis of the test.

    (ii) Write down the alternative hypothesis of the test.

[2]

(b) Write down the degree of freedom of the test.

[1]

(c) Find the value of χ^2_{calc}, the test statistic.

[2]

The critical value is given by 9.488.

(d) State the conclusion of the test with a reason.

[2]

Solution

(a) (i) H_0: The grade in Physics and the grade in Chemistry are independent.   A1   N1

    (ii) H_1: The grade in Physics and the grade in Chemistry are not independent.   A1   N1

[2]

(b) 4   A1   N1

[1]

(c) $\chi^2_{calc} = 8.968013758$ (A1) for correct value

       $\chi^2_{calc} = 8.97$ A1   N2

[2]

(d)   The null hypothesis is not rejected.   A1

       As $\chi^2_{calc} < 9.488$.   A1   N2

[2]

Exercise 97

1. A Mathematics teacher carried out an investigation of the results of the annual Mathematics examination. The examination paper consists of three sections: Algebra, Geometry and Statistics. A student can get a pass in a section if at least half of the answers are correct in that section.

The following table shows the number of students who passes or fails in each section:

		Sections		
		Algebra	Geometry	Statistics
Results	Pass	12	5	18
	Fail	8	15	2

A χ^2 test for independence is conducted at a 5% significance level.

(a)   (i)   Write down the null hypothesis of the test.

         (ii)   Write down the alternative hypothesis of the test.

[2]

(b)   Write down the degree of freedom of the test.

[1]

(c)   Find the value of χ^2_{calc}, the test statistic.

[2]

The critical value is given by 5.991.

(d)   State the conclusion of the test with a reason.

[2]

Your Practice Set – Applications and Interpretation for IBDP Mathematics

2. 200 children from three different countries are asked to fill in a survey to choose their most favourite fruits. The following table shows the distribution of the survey results:

		Most favourite fruits			
		Guava	Coconut	Banana	Pineapple
Country	Vietnam	21	20	25	4
	Malaysia	19	30	3	17
	Thailand	2	15	16	28

A χ^2 test for independence is conducted at a 1% significance level.

(a) (i) Write down the null hypothesis of the test.

    (ii) Write down the alternative hypothesis of the test.
[2]

(b) Write down the degree of freedom of the test.
[1]

(c) Find the p-value.
[2]

(d) State the conclusion of the test with a reason.
[2]

3. A survey is conducted among a sample of adults aged 18 or above, about their opinion on whether a new incinerator should be built in the town. The following table shows the distribution of the survey results:

		Preferences			Total
		Agree	Disagree	Neutral	
Age	18 to 30	2	23	5	30
	31 to 60	7	x	10	30
	61 or above	20	3	7	30
	Total	29	y	22	90

A χ^2 test for independence is conducted at a 5% significance level.

(a) Write down the null hypothesis of the test.
[1]

(b) Write down the values of x and y.
[2]

(c) Write down the degree of freedom of the test.
[1]

(d) Find the *p*-value.

[2]

(e) State the conclusion of the test with a reason.

[2]

4. The following table shows the distribution of the ages of staffs in an office and their number of investment bank accounts:

		Number of investment bank accounts					
		Zero	One	Two	Three	Four	**Total**
Age	18 to 37	*x*	4	5	3	1	**18**
	38 to 57	*y*	5	4	2	1	**16**
	58 to 77	11	2	1	1	1	**16**
	Total	**20**	**11**	**10**	**6**	**3**	**50**

A χ^2 test for independence is conducted at a 5% significance level.

(a) Write down the null hypothesis of the test.

[1]

(b) Write down the values of x and y.

[2]

(c) Write down the degree of freedom of the test.

[1]

(d) Find the *p*-value.

[2]

(e) State the conclusion of the test with a reason.

[2]

Your Practice Set – Applications and Interpretation for IBDP Mathematics

Paper 1 – *t*-test

Example

Robin purchased two samples of circular iron plates from two factories A and B. He wants to know whether the mean radius μ_A and μ_B of circular iron plates from Factory A and Factory B respectively are different. The following table shows the values of radii from the samples:

Radii of plates from Factory A (cm)	12.1	12.3	12.9	12.7			
Radii of plates from Factory B (cm)	11.9	11.7	11.1	11.0	11.0	12.0	12.5

A *t*-test is conducted at a 2% significance level.

(a) State an assumption of conducting the *t*-test. [1]

(b) (i) Write down the null hypothesis of the test.

(ii) Write down the alternative hypothesis of the test. [2]

(c) Find the *p*-value. [2]

(d) State the conclusion of the test with a reason. [2]

Solution

(a) The radii of plates are normally distributed.   A1   N1 [1]

(b) (i) $H_0: \mu_A = \mu_B$ A1   N1

(ii) $H_1: \mu_A \neq \mu_B$ A1   N1 [2]

(c) *p*-value = 0.0121929154   (A1) for correct value
    p-value = 0.0122   A1   N2 [2]

(d) The null hypothesis is rejected.   A1
    As *p*-value < 0.02.   A1   N2 [2]

Exercise 98

1. An environmental scientist catches certain fishes from two different ponds, one from Peru and another one from Indonesia. The lengths of fishes are recorded and studied, such that the mean lengths μ_1 and μ_2 of fishes from Peru and Indonesia respectively are compared. The following table shows the data:

Length of fishes from Peru (cm)	22.3	19.8	14.2	23.3	23.9	25.7	20.9
Length of fishes from Indonesia (cm)	15.2	16.3	21.1	18.4	19.1		

A t-test is conducted at a 5% significance level.

(a) State an assumption of conducting the t-test.

[1]

(b) (i) Write down the null hypothesis of the test.

(ii) Write down the alternative hypothesis of the test.

[2]

(c) Find the p-value.

[2]

(d) State the conclusion of the test with a reason.

[2]

2. Leo purchased two samples of bottles of milks from two suppliers, Dari and Luna. He wants to know whether the mean volume μ_D of bottles of milk from Dari is greater than the mean volume μ_L of bottles of milk from Luna. He measures the volumes precisely and summarize the data in the following table:

Volume of bottles of milk from Dari (mL)	330	329	330	332	327	329
Volume of bottles of milk from Luna (mL)	321	329	328	330	320	323

A t-test is conducted at a 5% significance level.

(a) State an assumption of conducting the t-test.

[1]

(b) (i) Write down the null hypothesis of the test.

(ii) Write down the alternative hypothesis of the test.

[2]

(c) Find the p-value.

[2]

(d) State the conclusion of the test with a reason.

[2]

3. The table below shows the typing speed (words per minute) of a group of students in a vocational training school before and after training:

Typing speed before training (words per minute)	38	32	37	40	15	18	29	23
Typing speed after training (words per minute)	40	46	56	51	30	20	37	40

The curriculum development officer of the school wants to investigate whether the students show improvement after training. Let μ_1 and μ_2 be the mean typing speeds of students before and after training respectively. A t-test is conducted at a 5% significance level, and it is assumed that the variances of the typing speeds for two data sets are the same.

(a) Write down the alternative hypothesis of the test.

[1]

(b) Find the p-value.

[2]

(c) Write down the value of t, the t-statistic.

[1]

(d) State the conclusion of the test with a reason.

[2]

4. Two surveys are conducted to measure the students' satisfaction on the services provided by the tuck shop. A score from 0 to 10 is used in the surveys, where 0 represents absolute dissatisfaction and 10 represents absolute satisfaction. The table below shows the results of the surveys:

Scores from the first survey	8	10	8	9	10	6	3	10
Scores from the second survey	10	10	7	10	9	6	5	

The student union committee members want to investigate whether the mean scores of the two surveys are different. Let μ_1 and μ_2 be the mean scores of the first survey and the second survey respectively. A t-test is conducted at a 10% significance level, and it is assumed that the variances of the scores from two surveys are the same.

(a) Write down the alternative hypothesis of the test. [1]

(b) Find the p-value. [2]

(c) Write down the value of t, the t-statistic. [1]

(d) State the conclusion of the test with a reason. [2]

Paper 2 – Problems Involving Chi-Squared Test

Example

The curriculum coordinator of a high school is going to investigate whether there is a relationship between the gender of a student and the choice of school overseas trip selected by the student. There are in total four choices of the overseas trip: Canada, USA, Mexico and Costa Rica.

80 students are asked to complete a survey about their preferences, and the following table shows the distribution of the results:

		Choices of overseas trips			
		Canada	USA	Mexico	Costa Rica
Gender	Male	10	22	4	4
	Female	7	12	8	13

A χ^2 test for independence is conducted at a 5% significance level.

(a) (i) Write down the null hypothesis of the test.

    (ii) Write down the alternative hypothesis of the test.

[2]

(b) Write down the degree of freedom of the test.

[1]

(c) Find the value of χ^2_{calc}, the test statistic.

[2]

The critical value is given by 7.815.

(d) State the conclusion of the test with a reason.

[2]

One student is selected at random.

(e) (i) Find the probability that the student is a male student and he chooses Costa Rica.

    (ii) Find the probability that the student chooses Canada.

    (iii) Given that the student chooses USA, find the probability that the student is a female.

[6]

(f) Two students are selected at random. Find the probability that both of them choose Mexico.

[3]

Solution

(a) (i) H_0: The gender of a student and the choices of overseas trips are independent. A1 N1

(ii) H_1: The gender of a student and the choices of overseas trips are not independent. A1 N1

[2]

(b) 3 A1 N1

[1]

(c) $\chi^2_{calc} = 9.568627451$ (A1) for correct value
$\chi^2_{calc} = 9.57$ A1 N2

[2]

(d) The null hypothesis is rejected. A1
As $\chi^2_{calc} > 7.815$. A1 N2

[2]

(e) (i) The required probability
$= \dfrac{4}{80}$ (A1) for correct formula
$= \dfrac{1}{20}$ A1 N2

(ii) The required probability
$= \dfrac{10+7}{80}$ (A1) for correct formula
$= \dfrac{17}{80}$ A1 N2

(iii) The required probability
$= \dfrac{12}{22+12}$ (A1) for correct formula
$= \dfrac{6}{17}$ A1 N2

[6]

(f) The required probability
$= \left(\dfrac{4+8}{80}\right)\left(\dfrac{4+8-1}{80-1}\right)$ (A2) for correct formula
$= \dfrac{33}{1580}$ A1 N3

[3]

Your Practice Set – Applications and Interpretation for IBDP Mathematics

Exercise 99

1. In order to improve the service, the manager of a bus company is investigating the number of on time arrivals and late arrivals of buses in four different bus stations. The manager wants to know whether there is an association between the punctuality of buses and the locations of bus stops.

    The following table shows the distribution of the 100 bus arrivals in different bus stations:

		Bus stations			
		A	B	C	D
Arrivals	On time	19	17	20	29
	Late	1	3	8	3

    A χ^2 test for independence is conducted at a 10% significance level.

    (a) (i) Write down the null hypothesis of the test.

        (ii) Write down the alternative hypothesis of the test.

    [2]

    (b) Write down the degree of freedom of the test.

    [1]

    (c) Find the value of χ^2_{calc}, the test statistic.

    [2]

    The critical value is given by 6.251.

    (d) State the conclusion of the test with a reason.

    [2]

    One bus arrival record is selected at random.

    (e) (i) Find the probability that the arrival record is on time and at the bus station C.

        (ii) Find the probability that the arrival record is late.

        (iii) Given that the arrival record is on time, find the probability that the arrival record is at the bus station A.

    [6]

    (f) Two bus arrival records are selected at random. Find the probability that both of them are on time.

    [3]

2. In an office, Hoang is recording the number of free lunches offered by the external sponsorships for staffs of different positions in the office in a particular working year. The following table shows the distribution of the results for 100 staffs:

		Positions			
		Administration	Customer Service	Human Resources	Others
Number of free lunches	0	4	4	20	6
	1	3	6	5	20
	2	2	10	2	2
	3 or more	1	10	3	2

A χ^2 test for independence is conducted at a 5% significance level.

(a) (i) Write down the null hypothesis of the test.

(ii) Write down the alternative hypothesis of the test.

[2]

(b) Write down the degree of freedom of the test.

[1]

(c) Find the p-value.

[2]

(d) Write down the value of χ^2_{calc}, the test statistic.

[1]

(e) State the conclusion of the test with a reason.

[2]

One staff is selected at random.

(f) (i) Find the probability that the staff is from the customer services department and is not offered any free lunch.

(ii) Find the probability that the staff is offered at least two free lunches.

(iii) Given that the staff is offered two free lunches, find the probability that the staff is from the administration department.

[6]

(g) Two staffs are selected at random. Find the probability that both of them are offered at most one free lunch.

[3]

Your Practice Set – Applications and Interpretation for IBDP Mathematics

3. The following table shows the nationality of 200 female students and X, the number of dolls they own:

		Number of dolls own			
		$1 \leq X \leq 5$	$6 \leq X \leq 10$	$11 \leq X \leq 15$	Total
Nationality	Colombia	35	15	13	63
	Ecuador	24	28	12	64
	Uruguay	41	17	15	73
	Total	a	b	40	200

(a) For the number of dolls own by students, write down

    (i) the class mark of the class $6 \leq X \leq 10$;

    (ii) the lower class boundary of the class $6 \leq X \leq 10$;

    (iii) the modal class.

[3]

(b) (i) Write down the values of a and b.

    (ii) Write down the estimated mean of X.

    (iii) Write down the estimated standard deviation of X.

[4]

(c) A female student is selected at random. Find the probability that she is from Colombia and owns at least six dolls.

[2]

A χ^2 test for independence is conducted at a 5% significance level.

(d) (i) Write down the null hypothesis of the test.

    (ii) Write down the alternative hypothesis of the test.

[2]

(e) Write down the degree of freedom of the test.

[1]

(f) Find the p-value.

[2]

(g) Write down the value of χ^2_{calc}, the test statistic.

[1]

(h) State the conclusion of the test with a reason.

[2]

4. The following table shows the ages of a certain number of interviewees participated in a survey and X, the number of yogurt parfait cups they eat in a particular month:

		Number of yogurt parfait cups ate			
		$1 \leq X \leq 10$	$11 \leq X \leq 20$	$21 \leq X \leq 30$	Total
Age	18 or below	50	75	125	250
	18 to 40	70	65	15	150
	41 to 60	55	10	5	70
	61 or above	25	4	1	a
	Total	200	154	b	n

(a) For the number of yogurt parfait cups eaten, write down

    (i) the class mark of the class $21 \leq X \leq 30$;

    (ii) the upper class boundary of the class $1 \leq X \leq 10$;

    (iii) the modal class.

[3]

(b) (i) Write down the values of a, b and n.

    (ii) Write down the exact value of the estimated mean of X.

    (iii) Write down the estimated standard deviation of X.

[5]

(c) An interviewee is selected at random. Find the probability that the interviewee is at most 40 years old, given that the interviewee eats at least 11 yogurt parfait cups in the month.

[3]

A χ^2 test for independence is conducted at a 1% significance level.

(d) (i) Write down the null hypothesis of the test.

    (ii) Write down the alternative hypothesis of the test.

[2]

(e) Write down the degree of freedom of the test.

[1]

(f) Find the p-value.

[2]

(g) Write down the value of χ^2_{calc}, the test statistic.

[1]

(h) State the conclusion of the test with a reason.

[2]

Your Practice Set – Applications and Interpretation for IBDP Mathematics

 Paper 2 – Problems Involving t-Test

Example

In a library, the habit of reading books of young people is investigated. The table below shows the number of fiction books read by two different groups of people in a particular month, after they have filled in a survey:

The group of 13-year-old students	5	6	5	4	7	9	5	3	2	4
The group of 23-year-old people	1	3	2	7	2	4	2	1		

The library manager wants to investigate whether 13-year-old students read more than 23-year-old people on average. Let μ_1 and μ_2 be the mean number of fiction books read by the 13-year-old students and the 23-year-old people respectively. A t-test is conducted at a 5% significance level.

(a) (i) Write down the null hypothesis of the test.

    (ii) Write down the alternative hypothesis of the test.

[2]

(b) Find the p-value.

[2]

(c) Write down the value of t, the t-statistic.

[1]

(d) State the conclusion of the test with a reason.

[2]

A person is randomly selected from the above 18 people.

(e) Given that the selected person reads at least four fiction books, find the probability that the person is 13 years old.

[2]

One person from each age group above is randomly selected.

(f) (i) Find the probability that both of them read less than four fiction books.

    (ii) Find the probability that at least one of them read less than four fiction books.

[4]

Solution

(a) (i) $H_0: \mu_1 = \mu_2$ A1   N1

    (ii) $H_1: \mu_1 > \mu_2$ A1   N1

[2]

(b) p-value $= 0.0153007379$ (A1) for correct value
p-value $= 0.0153$ A1   N2

[2]

(c) 2.38   A1   N1

[1]

(d) The null hypothesis is rejected.   A1
As p-value < 0.05.   A1   N2

[2]

(e) The required probability
$$= \frac{8}{8+2}$$ (A1) for correct formula
$$= \frac{4}{5}$$ A1   N2

[2]

(f) (i) The required probability
$$= \left(\frac{2}{10}\right)\left(\frac{6}{8}\right)$$ (A1) for correct formula
$$= \frac{3}{20}$$ A1   N2

    (ii) The required probability
$$= \left(\frac{2}{10}\right)\left(\frac{6}{8}\right) + \left(\frac{2}{10}\right)\left(\frac{2}{8}\right) + \left(\frac{8}{10}\right)\left(\frac{6}{8}\right)$$ (A1) for correct formula
$$= \frac{4}{5}$$ A1   N2

[4]

Your Practice Set – Applications and Interpretation for IBDP Mathematics

Exercise 100

1. A teacher wants to know whether the supplementary lessons between the mock examination and the official examination can improve students' mean score in the official examination. The table below shows the scores (out of 100) of some students in two examinations:

Mock exam score	47	53	85	76	80	65	68	90	77	
Official exam score	35	50	58	92	97	93	88	61	89	79

    Let μ_1 and μ_2 be the mean scores of the mock examination and the official examination respectively. A t-test is conducted at a 10% significance level, and it is assumed that the variances of the scores for two examinations are the same.

    (a) (i) Write down the null hypothesis of the test.

    (ii) Write down the alternative hypothesis of the test.
    [2]

    (b) Find the p-value.
    [2]

    (c) Write down the value of t, the t-statistic.
    [1]

    (d) State the conclusion of the test with a reason.
    [2]

    A score is randomly selected from the above 19 scores.

    (e) Given that the selected score is at least 80, find the probability that the score is from the mock examination.
    [2]

    One score from each examination is randomly selected.

    (f) (i) Find the probability that both of them are greater than 75.

    (ii) Find the probability that at most one of them are greater than 75.
    [4]

2. A social work is investigating whether organizing an anti-smoking event in public can decrease the mean smoking rates of citizens. The table below shows some samples of the number of times for a person to smoke in a day before the anti-smoking event and after the anti-smoking event:

Before anti-smoking event	12	8	15	7	3	10	16
After anti-smoking event	1	4	6	4	1		

Let μ_1 and μ_2 be the mean number of times for a person to smoke in a day before the anti-smoking event and after the anti-smoking event respectively. A t-test is conducted at a 1% significance level, and it is assumed that the variances of the scores for two examinations are the same.

(a) State an assumption of conducting the t-test.

[1]

(b) Write down the alternative hypothesis of the test.

[1]

(c) Find the p-value.

[2]

(d) Write down the value of t, the t-statistic.

[1]

(e) State the conclusion of the test with a reason.

[2]

A record is randomly selected from the above 12 records.

(f) (i) Find the probability that the score is from the after anti-smoking event and is an even number.

(ii) Given that the selected record is at least 4, find the probability that the score is from the after anti-smoking event.

[4]

Three records are randomly selected from the above 12 records.

(g) Find the probability that all of them are even numbers.

[3]

Your Practice Set – Applications and Interpretation for IBDP Mathematics

3. Samantha is a History teacher teaching the class A and the class B. She is comparing the performances of students in both classes in the History monthly assessments. The tables below show the scores (out of 100) of some students in both classes in the assessments in January and February:

	Class A								
Assessment in January	58	92	67	72	73	80	85	77	69
Assessment in February	90	32	97	45	62	60	50	74	

	Class B								
Assessment in January	65	68	58	57	70	49	79		
Assessment in February	*	*	*	*	*	*	*	*	*

Let μ_{A1} and μ_{B1} be the mean scores of the class A and class B respectively, in the assessment in January. The first t-test is conducted at a 5% significance level to test whether the mean scores of two classes are different.

(a) State an assumption of conducting the t-test.

[1]

(b) Write down the alternative hypothesis of this test.

[1]

(c) Find the p-value.

[2]

(d) Write down the value of t, the t-statistic.

[1]

(e) State the conclusion of the test with a reason.

[2]

Let μ_{A2} be the mean score of the class A in the assessment in February. The second t-test is conducted at a 5% significance level to test whether the mean score of the class A increases, and it is assumed that the variances of the scores for both assessments in class A are the same.

(f) (i) Write down the alternative hypothesis of this test.

(ii) Using the p-value approach to state the conclusion.

[4]

For the above scores of the assessment in January, one score from each class is randomly selected.

(g) Find the probability that only one of them is between 55 and 75 inclusive.

[3]

4. Robert is a football player who is attending a training session to practice his free kick skills since 2018. His coach is investigating whether the ball speed for Robert's free kick is increased in 2019. The tables below show the ball speeds (kilometre per hour) of some of his free kicks in 2018 and 2019:

Ball speed in 2018	88	85	74	99	101	95	65	108	90
Ball speed in 2019	103	88	94	97	105	132	127	79	

Let μ_{R1} and μ_{R2} be the mean ball speed of Robert's free kicks in 2018 and in 2019 respectively. The first t-test is conducted at a 10% significance level to test whether the mean ball speed for Robert's free kick has increased, and it is assumed that the variances of the ball speeds for Robert's free kick in both years are the same.

(a) State an assumption of conducting the t-test.

[1]

(b) Write down the alternative hypothesis of this test.

[1]

(c) Find the p-value.

[2]

(d) Write down the value of t, the t-statistic.

[1]

(e) State the conclusion of the test with a reason.

[2]

Marcel is Robert's friend who also attends the same free kick training program in 2019. Let μ_{M2} be the mean ball speed of Marcel's free kicks in 2019. The second t-test is conducted at a 10% significance level to test whether the mean ball speeds of their free kicks in 2019 are different. The tables below show the ball speeds (kilometre per hour) of some of Marcel's free kicks in 2019:

Ball speed in 2019	112	85	84	87	95	123	91	101	88

(f) (i) Write down the alternative hypothesis of this test.

(ii) Using the p-value approach to state the conclusion.

[4]

For the above ball speeds of free kicks in 2019, one record from each player is randomly selected.

(g) Find the probability that both of them are greater than 110 km/h or both of them are less than 90 km/h.

[3]

Your Practice Set – Applications and Interpretation for IBDP Mathematics

Answers

Chapter 1

Exercise 1

1.1 (a) 5.43×10^3 cm
    (b) 2.35×10^6 cm^2
1.2 (a) 1.52×10^4 cm
    (b) 2.49×10^7 cm^2
1.3 (a) 4.11×10^3 cm
    (b) 6.85×10^3 cm
1.4 (a) 8×10^4 cm
    (b) 8.04×10^4 cm

Chapter 2

Exercise 2

2.1 (a) 2.8125
    (b) 2.81
    (c) $2.805 \leq B < 2.815$
    (d) 0.978%
2.2 (a) 17.964 cm
    (b) The upper bound is 5.35 cm
        The lower bound is 5.25 cm
    (c) 0.200%
2.3 (a) 1007.881875 cm^3
    (b) The upper bound is 7.5 cm
        The lower bound is 6.5 cm
    (c) 0.0117%
2.4 (a) 37.5006 km
    (b) $37.495 \text{ km} \leq L < 37.505 \text{ km}$
    (c) 0.0197%

Chapter 3

Exercise 3

3.1 (a) (i) 3
        (ii) −4
    (b) Refer to solution
3.2 (a) (i) 3
        (ii) 4
    (b) Refer to solution
    (c) When the displacement of the particle is −2, its velocity is 0
3.3 (a) Refer to solution
    (b) −1
    (c) 0
    (d) −4
3.4 (a) Refer to solution
    (b) −4
    (c) 3
    (d) When the displacement of the car is 0, its velocity is 2

Exercise 4

4.1 (a) $x = -\dfrac{13}{5}$
    (b) $x = -3$
    (c) $y = -5$
4.2 (a) −3
    (b) $x = 1$
    (c) $y = 3$
4.3 (a) $x = 5$
    (b) $y = 1$
    (c) $\{x : x \neq 5\}$
    (d) $x < -0.317$ or $5 < x < 6.32$

4.4 (a) $x = \dfrac{5}{2}$

(b) $y = \dfrac{7}{2}$

(c) $\left\{y : y \neq \dfrac{7}{2}\right\}$

(d) $0.760 \leq x < \dfrac{5}{2}$ or $x \geq 7.24$

Exercise 5

5.1 (a) m represents the rate of change of the boiling point of water in degrees Celsius per 1 metre increase in vertical height above the sea level

(b) $m = -0.0036$

(c) 4440 m

5.2 (a) b represents the initial number of hotels

(b) $a = 10$, $b = 143$

(c) 223

5.3 (a) a represents the rate of change of the daily salary in dollars per 1 hour increase in working time

(b) b represents the fixed daily salary

(c) $a = 50$, $b = 200$

(d) $225

5.4 (a) p represents the rate of change of the area of the aluminium lamina in mm^2 per 1 degree Celsius increase in temperature

(b) q represents the area of the aluminium lamina at 0°C

(c) $p = 0.05$, $q = 5$

(d) 7 mm^2

Chapter 4

Exercise 6

6.1 (a) 2 and 4
(b) (i) $x = 3$
(ii) -1

6.2 (a) 1 and 10
(b) (i) $x = 5.5$
(ii) -20.25

6.3 (a) 0 and -7
(b) (i) $x = -3.5$
(ii) 24.5

6.4 (a) -3 and 3
(b) (i) $x = 0$
(ii) 13.5

Exercise 7

7.1 (a) $f(x) = (x-7)(x+5)$
(b) $x = -5$ and $x = 7$
(c) $\{y : y \geq -36\}$

7.2 (a) $f(x) = -2(x+1)(x+6)$
(b) $x = -1$ and $x = -6$
(c) $\{y : y \leq 12.5\}$

7.3 (a) $p = 5$ and $q = 11$
(b) $a = 1.5$
(c) $\{y : y \geq -13.5\}$

7.4 (a) $p = 0$ and $q = 18$
(b) $a = \dfrac{2}{3}$
(c) $\{y : y \leq 54\}$

Exercise 8

8.1 (a) $x = 2$
(b) $(2, 17)$
(c) $\{y : y \leq 17\}$

8.2 (a) 4
(b) $(7, -2)$
(c) $\{y : y \geq -2\}$

Your Practice Set – Applications and Interpretation for IBDP Mathematics

8.3 (a) $(-5, 12.5)$
    (b) $r = 10$, $a = -0.5$
8.4 (a) -5
    (b) $r = 4$, $a = -1$
    (c) t

Exercise 9

9.1 (a) -9920
    (b) $p = 108$
    (c) 100
9.2 (a) 5400
    (b) $p = 28$, $q = 40$
    (c) $(20, 5000)$
9.3 (a) 9840
    (b) $40 \leq x \leq 80$
    (c) 40
    (d) 60
9.4 (a) 20
    (b) $16 \leq x \leq 24$
    (c) 20

Exercise 10

10.1 (a) $A = 600 - 100x + 4x^2$
     (b) $x = 3.5$
     (c) 1046.5 cm^3
10.2 (a) $x^2 - 38x + 325 = 0$
     (b) $x = 25$
     (c) $\$20 / \text{cm}^2$
10.3 (a) $\$4.5$
     (b) $P = -15r^2 + 45r + 150$
     (c) $\$183.75$
10.4 (a) 12 m
     (b) 4 m
     (c) 6.90 m

Chapter 5

Exercise 11

11.1 (a) $a = 10$
     (b) $b = 1.5$
     (c) $y = 10$
11.2 (a) $p = 4$
     (b) $q = \dfrac{1}{2}$
     (c) $\{y : y > 7\}$
11.3 (a) $p = 2$, $q = -1$
     (b) $x = 0.792$
     (c) 0
11.4 (a) $a = 3$, $b = 4$
     (b) Increases
     (c) 1

Exercise 12

12.1 (a) $\$18000$
     (b) $\$9560$
     (c) 9.31 years
     (d) $\$2000$
12.2 (a) 270
     (b) 50
     (c) 48.6 days
12.3 (a) The initial price of the computer system
     (b) $A = 750$
     (c) 2.30 years
     (d) EUR 90
12.4 (a) The initial amount of electric charge stored
     (b) $p = 5$, $q = 0.8$
     (c) 1.60 hours

Exercise 13

13.1 (a) 12
     (b) 0.998 units
13.2 1000:1

13.3 (a) 0.903
(b) 1
(c) $x = -0.300$ or $x = 2.74$
13.4 4.28

Chapter 6

Exercise 14

14.1 (a) $a+b=10$, $9a+3b=42$
(or $36a+6b=120$)
(b) $a=2$, $b=8$
(c) $x=-2$
14.2 (a) $a=2$ and $b=-1$
(b) $c=-4$
14.3 (a) $a=2$ and $b=1$
(b) 1, 4
(c) $y=0$
14.4 (a) $p=10$, $q=6$, $r=1$
(b) 1, 4, 16
(c) $x=0$

Exercise 15

15.1 (a) $x+y=30$
(b) $3x+y=82$
(c) $x=26$, $y=4$
(d) 60
15.2 (a) $a+190b=8.25$
(b) $a+220b=9.21$
(c) $a=2.17$, $b=0.032$
(d) 240 °C
15.3 (a) $p=800$ and $q=16400$
(b) p represents the increase of the number of flats per year
(c) q represents the initial number of flats
15.4 (a) The price of one CD is USD 3.8
The price of one DVD is USD 5.5
(b) USD 18.4

Exercise 16

16.1 (a) (i) Refer to solution
(ii) Refer to solution
(iii) $36a+6b+c=928$
(b) $a=-2$, $b=0$ and $c=1000$
16.2 (a) (i) $x+y+z=8400$
(ii) Refer to solution
(b) $42x+84y+21z=655872$
(c) $x=800$, $y=7344$ and $z=256$
16.3 (a) (i) $\begin{cases} 10a+12b+13c=150 \\ 14a+8b+19c=178 \\ 22a+23b+7c=230 \end{cases}$
(ii) $a=5$, $b=4$ and $c=4$
(b) $290
16.4 (a) (i) $\begin{cases} 30x+16y=152 \\ 23x+15y+8z=114 \\ 11x+17y+18z=60 \end{cases}$
(ii) $x=4$, $y=2$ and $z=-1$
(b) A team drops 1 point for losing a game

Chapter 7

Exercise 17

17.1 (a) -7
(b) -141
(c) -1425
17.2 (a) 0.5
(b) 24
(c) 2037
17.3 (a) 3
(b) 6
(c) 105

Your Practice Set – Applications and Interpretation for IBDP Mathematics

17.4 (a) $-\dfrac{4}{3}$

(b) $-\dfrac{34}{3}$

(c) -960

Exercise 18

18.1 (a) 9.1 m
(b) 106 m

18.2 (a) $n = 20$
(b) 630

18.3 (a) $\$23700$
(b) 160 m

18.4 (a) 74
(b) $\$91800$

Exercise 19

19.1 (a) (i) $u_1 + 3d = 43$
(ii) $80u_1 + 3160d = -5320$
(iii) $u_1 = 52$, $d = -3$

(b) $u_n = -3n + 55$
(c) 18
(d) $S_n = -\dfrac{3}{2}n^2 + \dfrac{107}{2}n$
(e) $n = 75$

19.2 (a) (i) $u_1 + 10d = 17$
(ii) $96u_1 + 4560d = 4512$
(iii) $u_1 = 9$, $d = 0.8$

(b) $u_n = 0.8n + 8.2$
(c) 173
(d) 3177.6
(e) $n = 101$

19.3 (a) (i) $u_1 = 60$, $u_2 = 57$
(ii) -3

(b) 28
(c) (i) 60
(ii) $S_n = -\dfrac{3}{2}n^2 + \dfrac{123}{2}n$

(d) $n = 41$
(e) 165

19.4 (a) (i) $S_1 = \dfrac{5}{7}$, $S_2 = \dfrac{3}{2}$
(ii) $u_1 = \dfrac{5}{7}$, $u_2 = \dfrac{11}{14}$
(iii) $\dfrac{1}{14}$

(b) 45
(c) 306
(d) $u_n = \dfrac{1}{14}n + \dfrac{9}{14}$
(e) $n = 29$

Chapter 8

Exercise 20

20.1 (a) $r = \dfrac{1}{4}$
(b) $u_8 = \dfrac{1}{16}$
(c) $S_{12} = 1365$

20.2 (a) $r = \dfrac{4}{3}$
(b) $\sum_{n=1}^{7} u_n = 11217$
(c) $n = 8$

20.3 (a) $r = 1.25$
(b) 9
(c) 34.1

20.4 (a) $r = 1.6$
(b) $u_8 = 40.27$
(c) 7

Exercise 21

- 21.1
  - (a) 1.331 m
  - (b) 2.36 m
  - (c) 15.9 m
- 21.2
  - (a) $59
  - (b) $100 \times 0.9^{n-1}$
  - (c) $21527
- 21.3
  - (a) 308000 cm³
  - (b) 2
- 21.4
  - (a) 78.7 km
  - (b) 7.87 km
  - (c) 17th February

Exercise 22

- 22.1
  - (a) 1155 EUR
  - (b) 10648.9275 EUR
  - (c) 2018
  - (d) 2032
  - (e) 22050 EUR
- 22.2
  - (a) (i) v_n
    - (ii) t_n
    - (iii) u_n
    - (iv) w_n
  - (b) (i) $v_{100} = 99050$
    - (ii) 301250
  - (c) (i) 3200
    - (ii) 819150
  - (d) $m = 8$
- 22.3
  - (a) $(200n + 2200)$ m
  - (b) $x = 40$
  - (c) 6.24×10^1 km
  - (d) 3100 m
  - (e) $w = 43$
- 22.4
  - (a) 6120 EUR
  - (b) 40600 EUR
  - (c) 20500 EUR
  - (d) $m = 25$
  - (e) $n = 26$

Chapter 9

Exercise 23

- 23.1
  - (a) $P = 456800$ EUR
  - (b) $Q = 359500$ EUR
- 23.2
  - (a) $315000
  - (b) 10
- 23.3
  - (a) $P = 71300$
  - (b) 24
- 23.4 $t_1 - t_2 = 0.0861$

Exercise 24

- 24.1
  - (a) $P = 98000$ EUR
  - (b) $r = 6.03$
- 24.2
  - (a) $P = 860000$
  - (b) $n = 7.15$
- 24.3 $n = 3.96$
- 24.4 $k = 2.51$

Exercise 25

- 25.1
  - (a) $P = 229000$
  - (b) 5.4%
  - (c) $282000
- 25.2
  - (a) 10.9%
  - (b) 13000 EUR
- 25.3
  - (a) $r = 3.02$
  - (b) 0.983%
- 25.4
  - (a) $(9.5223 - i)\%$
  - (b) $i = 6.43$

Exercise 26

- 26.1
  - (a) $1290
  - (b) 16.7 years
- 26.2
  - (a) $64600
  - (b) $172000
- 26.3
  - (a) $26400
  - (b) $341

Your Practice Set – Applications and Interpretation for IBDP Mathematics

26.4   (a) $10800
       (b) $35200
       (c) $p = 144$

Exercise 27

27.1   (a)   (i) $16300
             (ii) $2350000
             (iii) $448000
       (b)   (i)   102 months
             (ii) $2135000
             (iii) $235000
       (c)   The option 1 is better
       (d)   The option 2 is better

27.2   (a)   (i) $891
             (ii) $2080
       (b)   (i)   56 months
             (ii) $4800
       (c)   The option 2 is better
       (d)   The option 1 is better
       (e) $r = 13.2$

27.3   (a)   (i) $91.9
             (ii) $1030
       (b)   (i)   96 months
             (ii) $986
             (iii)   The amount of interest paid in option 2 is less than that in option 1
       (c)   (i)   78 months
             (ii)   18 months

27.4   (a)   (i) $R_1 = 3000$
             (ii) $R_2 = 3060$
             (iii)   The difference between the total amounts to be paid for the version 1 and the version 2
             (iv)   Version 1
       (b)   (i) $R_3 = 253$
             (ii)   The amount of interest paid in version 3
             (iii)   The version 3 will have the smaller total amount to be paid

Chapter 10

Exercise 28

28.1   (a) $(-7, -24, 0)$
       (b)   (i) $(-7, -24, n)$
             (ii) $n = 60$

28.2   (a)   70
       (b) $(30, 55, 25)$
       (c)   (i) $(30, 55, 5)$
             (ii)   35

28.3   (a) $(-6, 14, -2)$
       (b)   41
       (c)   36.9°

28.4   (a) $h = -132$
       (b)   53.1°
       (c) $(-61, 116, 180)$

Exercise 29

29.1   (a) $x - 2y + 2 = 0$
       (b)   The x-intercept of L is -2
             The y-intercept of L is 1
       (c) $(-1, 0.5)$

29.2   (a) $3x + y + 20 = 0$
       (b)   The x-intercept of L is $-\dfrac{20}{3}$
             The y-intercept of L is -20
       (c) $\dfrac{1}{3}$

29.3   (a) $3x - y - 14 = 0$
       (b) L_1 and L_2 are parallel

29.4 (a) $5x - y + 20 = 0$
(b) L_1 and L_2 are perpendicular

Exercise 30

30.1 (a) The gradient of L_1 is $\dfrac{1}{2}$
The y-intercept of L_1 is 8
(b) $x - 2y + 12 = 0$

30.2 (a) $-\dfrac{3}{2}$
(b) $a = -4$
(c) $3x + 2y + 11 = 0$

30.3 (a) The gradient of L_1 is -3
The x-intercept of L_1 is -7
(b) $x - 3y + 7 = 0$

30.4 (a) (i) $\dfrac{1}{2}$
(ii) $-\dfrac{17}{4}$
(b) $8x + 4y + 17 = 0$
(c) $b = -5$

Exercise 31

31.1 (a) -3
(b) $3x + y + 6 = 0$
(c) $y = -3x + 6$
(d) $(-1, 3)$
(e) $x - 3y + 10 = 0$
(f) $k = 3$

31.2 (a) $k = 20$
(b) $\dfrac{1}{4}$
(c) $y = \dfrac{1}{4}x + 20$
(d) $y = -4x + 20$
(e) $4x + y - 80 = 0$
(f) $r = 16$
(g) $\left(\dfrac{240}{17}, \dfrac{400}{17}\right)$

31.3 (a) $-\dfrac{2}{3}$
(b) $2x + 3y - 6k = 0$
(c) $k = 10$
(d) $3x - 2y - 50 = 0$
(e) $h = 15$, $k = -2.5$
(f) $a = 0.1$
(g) 10 and 20

31.4 (a) $a = -4$, $b = 2$
(b) -16
(c) $h = -1$, $k = -18$
(d) 6
(e) $6x - y - 12 = 0$
(f) (i) $d = -96$
(ii) $d = \dfrac{8}{3}$

Exercise 32

32.1 (a) $-\dfrac{1}{2}$
(b) $x + 2y - 200 = 0$
(c) $m = 2$, $c = 0$
(d) $(40, 80)$
(e) 2000
(f) $r = 4$

Your Practice Set – Applications and Interpretation for IBDP Mathematics

32.2 (a) $\dfrac{1}{2}$

(b) $x - 2y + 60 = 0$

(c) $y = \dfrac{1}{2}x + 15$

(d) $y = -2x - 20$

(e) $(-14, 8)$

(f) 80

32.3 (a) $a = 40$

(b) $4x + 3y - 120 = 0$

(c) $y = \dfrac{3}{4}x - \dfrac{45}{2}$

(d) $c = 18$

(e) $CE = 45$

(f) 675

32.4 (a) $-\dfrac{1}{2}$

(b) $p = 20$, $q = -10$

(c) $y = -\dfrac{1}{2}x$

(d) $2x - y = 0$

(e) $OC = \sqrt{125}$

(f) $(5, 10)$

Chapter 11

Exercise 33

33.1 (a) $(0, 15)$

(b) 0

(c) $y = 15$

(d) The distance between A and D is equal to the distance between B and D

33.2 (a) (i) $(100, 0)$

(ii) $(0, 50)$

(iii) $(50, 25)$

(b) 2

(c) $y = 2x - 75$

33.3 (a) (i) $\dfrac{3}{4}$

(ii) $y = \dfrac{3}{4}x - 15$

(b) $(24, 3)$

(c) $(36, 12)$

33.4 (a) (i) $\dfrac{4}{3}$

(ii) $y = \dfrac{4}{3}x + 20$

(b) $\dfrac{4}{3}b + 20$

(c) (i) 25

(ii) $b = 18$

Exercise 34

34.1 (a) 27.5 km^2

(b) (i) $\dfrac{1}{2}$

(ii) $y = \dfrac{1}{2}x + \dfrac{9}{2}$

(c) So Yeon's home is on the boundary separating the Voronoi cells of the restaurant B and the restaurant C, which is equidistant to them

34.2 (a) (i) 3

(ii) $y = 3x - 8$

(b) The police station B

(c) $5 < x \leq 10$

34.3 (a) (i) 2.01

(ii) 2.15

(iii) 1.85

(b) The school C

34.4 (a) $(5, 5)$

(b) $2.5 < x < 5$

(c) Refer to solution

Exercise 35

35.1 (a) $x = 5$
(b) (i) $(7, 4)$
(ii) $k = 1$
(iii) $(5, 3)$

35.2 (a) (i) $(6, 6)$
(ii) $y = x$
(b) (i) $x = 5$
(ii) $(5, 5)$
(iii) 3.16

35.3 (a) (i) $(5, 5)$
(ii) $k = 10$
(b) $(1, 9)$
(c) Every position in the Voronoi cell of B has B to be the nearest supermarket

35.4 (a) (i) $x = 5$
(ii) $y = 5$
(b) (i) $(6, 6)$
(ii) $k = 7$
(iii) 2

Exercise 36

36.1 (a) (i) 3.125 km
(ii) 3 km
(iii) E
(b) 13 km

36.2 (a) (i) 2.02 km
(ii) 1.80 km
(iii) The hotel at E
(b) 1.34 km

36.3 (a) (i) 3.61 km
(ii) 3.80 km
(iii) Q
(b) The farm at B

36.4 (a) 4.40 km
(b) 4 km
(c) 17.6%

Chapter 12

Exercise 37

37.1 (a) $p = 4$
(b) $q = 2$
(c) $r = -2$

37.2 (a) $p = 16$
(b) $q = \dfrac{1}{4}$
(c) $r = 44$

37.3 (a) $p = 2\pi$
(b) $q = 6$
(c) $r = 15°$

37.4 (a) $p = 10$
(b) $q = \dfrac{1}{3}$
(c) $r = -1080°$

Exercise 38

38.1 (a) (i) 3.5
(ii) 2
(b) Refer to solution

38.2 (a) (i) 3
(ii) 2
(b) Refer to solution

38.3 (a) (i) 4
(ii) $360°$
(b) Refer to solution

38.4 (a) (i) 3
(ii) $720°$
(b) Refer to solution

Exercise 39

39.1 (a) $180°$
(b) -2
(c) $x = 45°$ or $x = 225°$

Your Practice Set – Applications and Interpretation for IBDP Mathematics

39.2 (a) 3
 (b) 4
 (c) $x = 0°$, $x = 360°$ or $x = 720°$

39.3 (a) 90°
 (b) $\{y : 1 \leq y \leq 5\}$
 (c) $x = -82.5°$, $x = -52.5°$, $x = 7.5°$ or $x = 37.5°$

39.4 (a) $\dfrac{1}{2}$
 (b) $\left\{y : -\dfrac{3}{4} \leq y \leq \dfrac{1}{4}\right\}$
 (c) $x = 180°$ or $x = 900°$

Exercise 40

40.1 (a) (i) 5.5 hours
 (ii) 1.4 m
 (b) (i) 0.7
 (ii) $\left(\dfrac{360}{11}\right)°$
 (iii) 1.1
 (c) 16:15

40.2 (a) (i) 13 hours
 (ii) 2.4 m
 (b) (i) 1.2
 (ii) $\left(\dfrac{360}{13}\right)°$
 (iii) 3
 (c) 16:00

40.3 (a) 9:18
 (b) (i) 45
 (ii) $10°$
 (iii) 46
 (c) 10:22

40.4 (a) (i) 35
 (ii) $\left(\dfrac{180}{13}\right)°$
 (iii) 38
 (b) 18.1 m
 (c) 14:00

Chapter 13

Exercise 41

41.1 (a) $AB = 8.13$ cm
 (b) $BC = 3.38$ cm

41.2 (a) $AB = 21.8$ cm
 (b) 228 cm^2

41.3 (a) $BC = 68.7$ cm
 (b) $AB = 55.4$ cm

41.4 (a) $B\hat{A}C = 70.7°$
 (b) $BC = 53.8$ cm

Exercise 42

42.1 (a) $A\hat{P}B = 101°$
 (b) $PB = 86.4$ m
 (c) $PC = 48.3$ m

42.2 (a) $P\hat{R}Q = 45°$
 (b) $Q\hat{P}R = 21°$
 (c) $PQ = 9.87$ m

42.3 (a) $78.6°$
 (b) $x = 1.58$
 (c) $h = 7.84$

42.4 (a) $29°$
 (b) $38.3°$
 (c) 18.6 m

Exercise 43

43.1 (a) $60°$
 (b) $BC = 9$ cm
 (c) 3π cm

43.2 (a) $A\hat{B}C = 120°$
 (b) $60°$
 (c) 47.8 cm

43.3 (a) $AC = 14$ cm
 (b) 119 cm^2

43.4 (a) AC = 12 cm
(b) AB = √112 cm
(c) 37.4 cm

Exercise 44

44.1 (a) 197 cm
(b) 307 cm
(c) 5410 cm²
44.2 (a) 59.0 cm
(b) 190 cm²
44.3 (a) 63.0
(b) 192 cm²
44.4 (a) 3.74 cm
(b) 30.0 cm²

Exercise 45

45.1 (a) 19400 cm²
(b) 4810 cm²
(c) 14600 cm²
45.2 (a) 2430 cm
(b) 2240 cm
(c) 4670 cm
45.3 (a) AÔB = 106°
(b) 371 cm²
(c) 179 cm²
45.4 (a) AÔB = 97.2°
(b) 67.8 cm²
(c) 243 cm

Exercise 46

46.1 (a) 257°
(b) (i) AÊC = 26°
(ii) AE = 1780 km
(c) DE = 1400 km
(d) (i) BE = 1380 km
(ii) 61.2 km/h
46.2 (a) AC = 16.6 km
(b) 98.2 km²
(c) (i) 196 km²
(ii) BC = 28.5 km
(d) (i) DC = 13.9 km
(ii) BD = 36.5 km
(iii) 1.28 hours
46.3 (a) AC = 44.9 km
(b) 1020 km²
(c) $\theta° = 54.9°$
(d) Speed of Q = 47.0 km/h
46.4 (a) BD = 52.4 km
(b) 1830 km²
(c) $\theta° = 104°$
(d) 2 hours 33 minutes

Exercise 47

47.1 (a) AĈB = 47.9°
(b) (i) BÂC = 60.6°
(ii) BC = 24.4 km
(c) (i) 238°
(ii) 408 cm²
47.2 (a) AC = 35.2 cm
(b) BC = 39.8 cm
(c) (i) BÔC = 120°
(ii) 553 cm²
(iii) 324 cm²
47.3 (a) AC = 19.7 cm
(b) OC = 26.8 cm
(c) 52.7 cm²
47.4 (a) AĈB = 32.1°
(b) OC = 52.5 cm
(c) 102 cm

Your Practice Set – Applications and Interpretation for IBDP Mathematics

Chapter 14

Exercise 48

48.1 (a) 3 cm
(b) 12π cm^3
(c) 75.4 cm^2

48.2 (a) 3.43 cm
(b) 84.7 cm^3
(c) 111 cm^2

48.3 (a) 3.32 cm
(b) 13.4 cm
(c) 140 cm^2

48.4 (a) 9 cm
(b) 40 cm
(c) 1080π cm^3

Exercise 49

49.1 (a) 4670 m^3
(b) 754 m^2

49.2 (a) 1470 cm^3
(b) 653 cm^2

49.3 (a) 30 m
(b) 5650 m^2

49.4 (a) 2 mm
(b) 71.2 mm^3

Exercise 50

50.1 (a) 2.23×10^4 cm^3
(b) 16.5 cm

50.2 (a) 7.37×10^3 cm^3
(b) 12.1 cm

50.3 (a) 4900π cm^3
(b) 9.02×10^1 cm

50.4 (a) 2.06×10^4 cm^3
(b) 3:1

Exercise 51

51.1 (a) 2830 cm^2
(b) 17.1 cm

51.2 (a) 2460 cm^2
(b) 23.3%

51.3 (a) 2940 cm^2
(b) 19.6%

51.4 (a) 704 cm^2
(b) 47.7%

Exercise 52

52.1 (a) 2090 cm^3
(b) 942 cm^2
(c) 22.2 cm
(d) 81.2°
(e) 2880 cm^2

52.2 (a) 26000 cm^3
(b) 6720 cm^2
(c) 44.2 cm
(d) 61.1°
(e) 52.0°

52.3 (a) 22.6°
(b) 18100 cm^2
(c) 161000 cm^3
(d) $R = \dfrac{5}{12} H$
(e) $H = 76.2$ cm, $R = 31.7$ cm

52.4 (a) 36.9°
(b) 59.4 cm
(c) 65856 cm^3
(d) 1:0.630

Chapter 15

Exercise 53

53.1 (a) $f'(x) = 12x^3 + \dfrac{1}{2}$

  (b) $\dfrac{1}{2}$

  (c) $\dfrac{2}{23}$

53.2 (a) $f'(x) = 8x$

  (b) 2

  (c) $a = -2$

53.3 (a) $f(-2) = \dfrac{5}{2}$

  (b) $f'(x) = x + \dfrac{1}{x^2}$

  (c) $a = 3$

53.4 (a) 0

  (b) $f'(x) = \dfrac{3}{a}x^2 + 1$

  (c) $a = 6$

Exercise 54

54.1 (a) $f'(x) = 24x^3 - 42x$

  (b) 108

  (c) $y = 108x - 204$

54.2 (a) $f'(x) = 3 + 8x^{-3}$

  (b) $-\dfrac{1}{11}$

  (c) $y = -\dfrac{1}{11}x - \dfrac{10}{11}$

54.3 (a) $f'(x) = 3ax^2 - 4x$

  (b) $a = 4$

  (c) $y = 96x - 197$

54.4 (a) $f'(x) = -3ax^2$

  (b) $\dfrac{1}{12a}$

  (c) -1

  (d) $y = -\dfrac{1}{12}x + \dfrac{67}{6}$

Exercise 55

55.1 (a) $f'(x) = 4x - 1$

  (b) $g'(x) = 2x$

  (c) $x = \dfrac{1}{2}$

  (d) 1

55.2 (a) $f'(x) = 3x^2 + 2x$

  (b) $g'(x) = 1$

  (c) $x = -1$ or $x = \dfrac{1}{3}$

  (d) $x = -\dfrac{1}{3}$

55.3 (a) $f'(x) = 24 - 3x^2$

  (b) $g'(x) = 3x^2$

  (c) $x = -2$ or $x = 2$

  (d) 4

55.4 (a) (i) $f'(x) = -2ax$

    (ii) $g'(x) = -4$

  (b) $a = 1$

  (c) 1

  (d) $\dfrac{9}{4}$

Exercise 56

56.1 (a) $f'(x) = 6x^2 - 66x + 108$

  (b) $2 < x < 9$

56.2 (a) $f'(x) = -4x^3 + 60x^2 - 284x + 420$

  (b) $x < 3$ or $5 < x < 7$

Your Practice Set – Applications and Interpretation for IBDP Mathematics

56.3 (a) (i) $y = 5$
 (ii) $y = 2$
 (b) $x < 3$ or $x > 11$
 (c) $(11, 2)$
 (d) $f(9)$

56.4 (a) (i) $y = 0$
 (ii) $y = 12$
 (b) $x < -4$, $1 < x < 5$ or $x > 5$
 (c) $(1, 20)$
 (d) $f(2)$

Exercise 57

57.1 (a) $C'(x) = 2x - \dfrac{54}{x^2}$
 (b) 3 kg
 (c) \$33

57.2 (a) 1000
 (b) $P'(t) = 3t^2 - 24t + 36$
 (c) 2
 (d) 32001000

57.3 (a) $t = 2.12$ or $t = 4.10$
 (b) $Q'(t) = 8t - \dfrac{216}{t^2}$
 (c) $t = 3$

57.4 (a) $P'(t) = -3t^2 + 18t - 24$
 (b) $t = 2$
 (c) The minimum price of the share is \$700, which is greater than \$690

Exercise 58

58.1 (a) $r = -29$
 (b) $f'(x) = -3x^2 + 6x + 24$
 (c) $(4, 79)$
 (d) $x < -2$ or $x > 4$
 (e) (i) -1
 (ii) $f'(0) = 24$
 (iii) $24x - y - 1 = 0$
 (iv) $x = \dfrac{1}{24}$

58.2 (a) $r = -5$
 (b) $f'(x) = 6x^2 - 150$
 (c) $(5, -650)$
 (d) $x < -5$ or $x > 5$
 (e) (i) -2
 (ii) $f'(-1) = -144$
 (iii) $\dfrac{1}{144}$
 (iv) $y = \dfrac{1}{144}x - \dfrac{287}{144}$
 (v) $x = 287$

58.3 (a) $x = 0$
 (b) $f'(x) = 125 - \dfrac{64}{x^3}$
 (c) $(0.8, 150)$
 (d) $0 < x < 0.8$
 (e) (i) 157
 (ii) $f'(1) = 61$
 (iii) $61x - y + 96 = 0$
 (f) $y < 150$

58.4
(a) $x = 0$
(b) $f'(x) = x - \dfrac{16}{x^3}$
(c) $(2, 4)$
(d) $-1 < x < 0$ or $x > 2$
(e) (i) 8.5
  (ii) $f'(4) = \dfrac{15}{4}$
  (iii) $-\dfrac{4}{15}$
  (iv) $y = -\dfrac{4}{15}x + \dfrac{287}{30}$
(f) $y < 4$

Exercise 59

59.1
(a) (i) $h = \dfrac{27 - r^2}{2r}$
  (ii) Refer to solution
(b) $\dfrac{dV}{dr} = \dfrac{27}{2}\pi - \dfrac{3}{2}\pi r^2$
(c) 3 cm
(d) (i) 27π cm^3
  (ii) $0 \leq V \leq 27\pi$
(e) 1.17 cm and 4.51 cm

59.2
(a) (i) $(64 - 2x)$ cm
  (ii) Refer to solution
(b) $\dfrac{dV}{dx} = 12x^2 - 512x + 4096$
(c) $x = \dfrac{32}{3}$
(d) (i) 19418 cm^3
  (ii) $0 \leq V \leq 19418$
(e) 7280 cm^2

59.3
(a) (i) Refer to solution
  (ii) Refer to solution
(b) $\dfrac{dV}{dr} = \dfrac{25\sqrt{3}}{3} - \dfrac{3}{8}r^2$
(c) $r = 6.20$
(d) 59.7 cm^3
(e) 1.51×10^{-1} kg/cm^3

59.4
(a) (i) $h = \dfrac{672}{\pi r^2}$
  (ii) Refer to solution
(b) $\dfrac{dA}{dr} = \pi r - \left(\dfrac{1344}{\pi} + 336\right)\dfrac{1}{r^2}$
(c) $r = 6.24$
(d) 184 cm^2
(e) 8 buckets

Chapter 16

Exercise 60

60.1 $f(x) = 2x^3 - x^2 - 8x + 7$

60.2
(a) $f(x) = \dfrac{1}{3}x^3 - 36x + \dfrac{3}{x} + 98$
(b) $a = 1.5$

60.3
(a) $f(x) = \dfrac{1}{3}x^3 - 32x^2 + 12x + 1008$
(b) 1008
(c) 3024

60.4
(a) $f(x) = \dfrac{100}{x^2} + \dfrac{20}{x} + 1$
(b) $b = 20$

Exercise 61

61.1
(a) (i) 10
  (ii) $(7.67, 50.8)$
(b) (i) $\displaystyle\int_3^{10}\left(\begin{array}{c}-x^3 + 16x^2 \\ -69x + 90\end{array}\right)dx$
  (ii) $\dfrac{2401}{12}$

61.2
(a) (i) -5
  (ii) $(5, 0)$
(b) (i) $\displaystyle\int_{-5}^{0}\left(\begin{array}{c}x^3 - 5x^2 \\ -25x + 125\end{array}\right)dx$
  (ii) $\dfrac{6875}{12}$

Your Practice Set – Applications and Interpretation for IBDP Mathematics

61.3 (a) (i) $a = -7$, $b = 3$
        (ii) $(-2, 50)$
   (b) $c = -6$
61.4 (a) (i) $a = 20$, $b = 40$
        (ii) $\int_{20}^{40} \begin{pmatrix} -2x^2 + 120x \\ -1600 \end{pmatrix} dx$
        (iii) $\dfrac{8000}{3}$
   (b) $c = \dfrac{170}{3}$

Exercise 62

62.1 (a) (i) $\int_{-2}^{10} \begin{pmatrix} 1.5(x+2) \\ (x-10)^2 \end{pmatrix} dx$
        (ii) 2592
   (b) 12960

62.2 (a) 36π
   (b) 3

62.3 (a) (i) $\int_{4}^{8} (x-4)^2 (8-x) dx$
        (ii) $\dfrac{64}{3}$
   (b) $h = 15$

62.4 (a) $\dfrac{8192}{15}$
   (b) $h = 1.125$

Exercise 63

63.1 (a) 1.5
   (b) 8.84
63.2 (a) $n = 3$
   (b) 0.745
63.3 (a) $a = 3.6$
   (b) 372
63.4 (a) $\int_{2}^{b} \dfrac{1}{2x} dx$
   (b) $b = 4$
   (c) 0.347

Exercise 64

64.1 (a) 6.06
   (b) 0.991%
64.2 (a) 0.6
   (b) 78.6
   (c) 2.98%
64.3 (a) (i) $\dfrac{5}{3}$
        (ii) $f(5.75) = \dfrac{14}{3}$, $f(5.875) = \dfrac{19}{3}$
   (b) 5.5
   (c) Overestimate
64.4 (a) $a = 0.35$, $b = 0.5$
   (b) 0.162
   (c) Overestimate

Chapter 17

Exercise 65

65.1 (a) 7
   (b) 7
   (c) 9
   (d) 5.51
65.2 (a) 6.5
   (b) (i) 7
       (ii) 8.5
       (iii) 4.5
       (iv) 4
65.3 (a) $x = 8$
   (b) 6, 8
   (c) 3
65.4 (a) (i) 48
       (ii) 59
       (iii) $x = 6$
   (b) 19
   (c) 6.29

Exercise 66

66.1 (a) (i) 27.5
       (ii) $30 \leq x < 35$
   (b) (i) 34.625
       (ii) 6.06
       (iii) 36.7

66.2 (a) $f = 15$
   (b) (i) Continuous
       (ii) 175 USD
       (iii) $0 \leq x < 50$
   (c) (i) 81.75 USD
       (ii) 67.9 USD

66.3 (a) 12
   (b) (i) 18
       (ii) $12 \leq x < 16$
   (c) (i) 11.1
       (ii) 4.95
   (d) 16

66.4 (a) $p = 6$, $q = 5$, $r = 3$, $s = 2$
   (b) $0 \leq x < 3$
   (c) (i) 2.87
       (ii) 3.09
       (iii) 7.49%

Exercise 67

67.1 (a) (i) 21
       (ii) 18
   (b) 2.54
   (c) 1.15

67.2 (a) (i) 90
       (ii) 50
   (b) 24.5
   (c) 2.60

67.3 (a) 12
   (b) 50
   (c) Discrete

67.4 (a) 6
   (b) 24.5

Exercise 68

68.1 (a) $a = 3$, $b = 14$
   (b) 24

68.2 (a) $a = 63$, $b = 73$
   (b) 89

68.3 (a) (i) 34
       (ii) 24
       (iii) 12
   (b) 5

68.4 (a) (i) 5
       (ii) 8
       (iii) 6
   (b) 9

Exercise 69

69.1 (a) 10
   (b) (i) 30
       (ii) 72

69.2 (a) 108
   (b) (i) 19
       (ii) 1.5

69.3 (a) 21
   (b) (i) 40
       (ii) 44

69.4 (a) 10
   (b) (i) 19
       (ii) 34

Exercise 70

70.1 (a) (i) $7.5
       (ii) 20
   (b) (i) 75
       (ii) 125
   (c) 70
   (d) 50
   (e) Simple random sampling

Your Practice Set – Applications and Interpretation for IBDP Mathematics

70.2 (a) (i) 1.5 cm
   (ii) 20
   (iii) 40%
   (iv) 1
(b) $90
(c) 80

70.3 (a) 25 minutes
(b) 15 minutes
(c) 60
(d) 40
(e) 52.5
(f) Systematic sampling

70.4 (a) 35 minutes
(b) 10 minutes
(c) 30
(d) 25
(e) 55
(f) $\dfrac{1}{16}$

Chapter 18

Exercise 71

71.1 (a) (i) $\dfrac{3}{5}$
   (ii) $\dfrac{8}{11}$
(b) $\dfrac{11}{38}$

71.2 (a) (i) $\dfrac{3}{5}$
   (ii) $\dfrac{9}{14}$
(b) $\dfrac{3}{35}$

71.3 (a) (i) $\dfrac{18}{25}$
   (ii) $\dfrac{5}{8}$
(b) $\dfrac{7}{100}$

71.4 (a) (i) $a = 4$, $b = 6$
   (ii) $\dfrac{6}{7}$
(b) $\dfrac{1}{330}$

Exercise 72

72.1 (a) (i) 4
   (ii) 5
(b) $\dfrac{2}{15}$

72.2 (a) (i) 2
   (ii) 13
(b) $\dfrac{1}{20}$

72.3 (a) (i) 0.4
   (ii) 0.2
(b) 0.7

72.4 (a) (i) 0.3
   (ii) 0.1
(b) 0.1

Exercise 73

73.1 (a) Refer to solution
(b) $\dfrac{15}{28}$

73.2 (a) Refer to solution
(b) $\dfrac{5}{6}$

73.3 (a) $\dfrac{5}{8}$
(b) $\dfrac{13}{40}$
(c) $\dfrac{3}{13}$

73.4 (a) $\dfrac{3}{5}$
(b) $\dfrac{2}{3}$
(c) $\dfrac{3}{5}$

Exercise 74

- **74.1** (a) 0.2
  (b) 0.4
  (c) 0.52
- **74.2** (a) 0.15
  (b) 0.5
  (c) 0.65
- **74.3** (a) 0.12
  (b) 0.58
- **74.4** (a) 0.3
  (b) 0.79
  (c) 0.21

Exercise 75

- **75.1** (a) $6k^4$
  (b) 0.2
  (c) 0.0704
  (d) 0.88
- **75.2** (a) $4k^4$
  (b) $\dfrac{1}{10}$
  (c) $\dfrac{259}{2500}$
- **75.3** (a) Refer to solution
  (b) Refer to solution
  (c) $\dfrac{1}{3}$
  (d) 1
- **75.4** (a) Refer to solution
  (b) $P(B) = \dfrac{9}{10}$

Exercise 76

- **76.1** (a) 0.05
  (b) 0.6
  (c) (i) 0.546
      (ii) 0.154
      (iii) 0.6
      (iv) 0.257
- **76.2** (a) 25%
  (b) 75%
  (c) (i) 0.4536
      (ii) 0.0964
      (iii) 0.37
      (iv) 0.261
- **76.3** (a) 30%
  (b) 15%
  (c) 70%
  (d) (i) 0.353
      (ii) 0.214
  (e) (i) 54%
      (ii) 31%
- **76.4** (a) 10%
  (b) 40%
  (c) 75%
  (d) (i) 0.533
      (ii) 0.2
  (e) (i) 38%
      (ii) 12%

Chapter 19

Exercise 77

- **77.1** (a) 0.05
  (b) 1.45
- **77.2** (a) $\dfrac{3}{10}$
  (b) 38
- **77.3** (a) 3
  (b) $\dfrac{18}{7}$
  (c) $\dfrac{2}{49}$
- **77.4** (a) 1
  (b) $\dfrac{5}{2}$
  (c) $\dfrac{1}{8}$

Your Practice Set – Applications and Interpretation for IBDP Mathematics

Exercise 78

78.1 (a) $\dfrac{1}{6}$

(b) $\dfrac{10}{13}$

78.2 (a) $\dfrac{1}{5}$

(b) $\dfrac{7}{20}$

78.3 (a) $\dfrac{1}{7}$

(b) $\dfrac{14}{19}$

78.4 (a) $\dfrac{1}{7}$

(b) $\dfrac{9}{10}$

Exercise 79

79.1 (a) $a+b=0.5$

(b) $3a+4b=1.82$

(c) $a=0.18$, $b=0.32$

79.2 (a) $a+b=0.8$

(b) $30a+40b=26$

(c) $a=0.6$, $b=0.2$

79.3 (a) $a=0.4$

(b) $b=0.3$ and $c=0.2$

(c) $P(Y=50)=0.12$

79.4 (a) $c=0.3$

(b) $a=0.1$ and $b=0.2$

(c) $P(Y=36)=0.04$

Exercise 80

80.1 (a) (i) 0.36

(ii) 0.6

(b) (i) 0.16

(ii) 0.6

(c) Refer to solution

(d) 5.6

80.2 (a) (i) 0.06

(ii) 0.54

(b) (i) 0.48

(ii) $\dfrac{7}{23}$

(c) Refer to solution

(d) 16.45

80.3 (a) (i) $\dfrac{7}{20}$

(ii) $\dfrac{2}{5}$

(b) (i) $\dfrac{1}{20}$

(ii) $\dfrac{7}{10}$

(c) Refer to solution

(d) $225

80.4 (a) (i) 0.2

(ii) 0.6

(b) (i) 0.1

(ii) $\dfrac{2}{3}$

(c) Refer to solution

(d) $4.8

Exercise 81

81.1 (a) (i) $\dfrac{1}{9}$

(ii) $\dfrac{1}{6}$

(iii) $\dfrac{3}{10}$

(b) (i) $\dfrac{13}{18}$

(ii) $\dfrac{12}{13}$

81.2 (a) (i) $\dfrac{5}{36}$

(ii) $\dfrac{5}{18}$

(iii) $\dfrac{3}{5}$

(b) (i) $\dfrac{7}{12}$

(ii) 3.2

81.3 (a) (i) $\dfrac{1}{9}$

(ii) $\dfrac{5}{9}$

(iii) $\dfrac{2}{5}$

(b) 9

81.4 (a) (i) $\dfrac{1}{9}$

(ii) $\dfrac{2}{9}$

(iii) $\dfrac{5}{6}$

(b) 72

Chapter 20

Exercise 82

82.1 (a) 2.5
(b) 1.875
(c) 0.250
82.2 (a) 30
(b) 72
(c) 0.140
82.3 (a) 0.8
(b) 16
(c) 0.205
82.4 (a) 0.45
(b) 4000
(c) 2.3511×10^{-11}

Exercise 83

83.1 (a) 4.8
(b) 0.0131
(c) 0.0000749
83.2 (a) 16.2
(b) 0.0598
(c) 0.455
83.3 (a) 1
(b) 5.60×10^{-7}
(c) 0.922
83.4 (a) 6.21
(b) 0.270
(c) 0.00527

Exercise 84

84.1 (a) $\binom{120}{3} p^3 (1-p)^{117}$
(b) 0.0149 or 0.0388
84.2 (a) $5p^4(1-p)$
(b) 0.638 or 0.914
84.3 (a) $10q^9(1-q) + q^{10}$
(b) 0.654
84.4 (a) $(1-q)^{100} + 100q(1-q)^{99}$
(b) 0.0524

Exercise 85

85.1 (a) 0.4016
(b) 0.833
(c) 0.310
(d) 3
85.2 (a) 0.26
(b) 0.692
(c) 0.222
(d) 10
85.3 (a) 0.383
(b) 0.847
(c) 0.285
(d) 10

Your Practice Set – Applications and Interpretation for IBDP Mathematics

85.4   (a) $0.48 - 0.18p$

(b) $\dfrac{0.3p}{0.48 - 0.18p}$

(c)   0.0286

(d)   20

Chapter 21

Exercise 86

86.1 (a) 0.28
     (b) 0.22
     (c) 0.22

86.2 (a) 0.15
     (b) 0.35
     (c) 0.7

86.3 (a) 0.07
     (b) 0.43
     (c) 0.93

86.4 (a) $\dfrac{5}{6}$

     (b) $\dfrac{9}{11}$

     (c) $\dfrac{2}{3}$

Exercise 87

87.1 (a) Refer to solution
     (b) 0.0228
     (c) 61.8

87.2 (a) Refer to solution
     (b) 0.2654
     (c) 4.76

87.3 (a) Refer to solution
     (b) 0.4479
     (c) 32.1

87.4 (a) Refer to solution
     (b) 0.3697
     (c) 158

Exercise 88

88.1 (a) 230
     (b) 250
     (c) 20.2

88.2 (a) 156
     (b) 164
     (c) 8.39

88.3   10.3

88.4   2.58

Exercise 89

89.1 (a) (i) 0.00839
         (ii) 0.0301
     (b) 0.0000705
     (c) (i) 0.839
         (ii) 0.0525

89.2 (a) (i) 0.00621
         (ii) 0.931
     (b) 0.0123
     (c) (i) 0.373
         (ii) 0.994

89.3 (a) (i) 66.9
         (ii) 0.319
     (b) 0.255
     (c) (i) 3.1875
         (ii) 0.529

89.4 (a) (i) 9.51
         (ii) 0.773
     (b) 0.001
     (c) (i) 4.68
         (ii) 0.00149

Chapter 22

Exercise 90

90.1 (a) Refer to solution
     (b) Refer to solution
     (c) Strong, negative
     (d) 77

90.2	(a)	Refer to solution
	(b)	(70, 70)
	(c)	Refer to solution
	(d)	Refer to solution
90.3	(a)	Refer to solution
	(b)	Refer to solution
	(c)	$p = 65$, $q = 95$
90.4	(a)	Refer to solution
	(b)	15
	(c)	$p = 20$, $q = 40$

Exercise 91

91.1	(a)	(i)	$a = 0.2$, $b = 52.4$
		(ii)	69.4
	(b)	(i)	0.183
		(ii)	Weak, Positive
91.2	(a)	(i)	$a = -2.09$, $b = 96.1$
		(ii)	77.3°C
	(b)	(i)	−0.607
		(ii)	Moderate, Negative
91.3	(a)	(i)	$a = 0.712$, $b = 7.22$
		(ii)	25.0
	(b)		−0.989
	(c)		0.54
91.4	(a)	(i)	−0.957
		(ii)	$a = -0.746$, $b = 6.75$
	(b)		0.178
	(c)		1.431 hours

Exercise 92

92.1	(a)	(i)	$a = 0.157$, $b = -5.75$
		(ii)	a represents the average increase of university entrance mark when the public exam score is increased by 1
	(b)		22.4
92.2	(a)	(i)	$a = 3.42$, $b = 1.55$
		(ii)	b represents the expected sales in 2011
	(b)		10.1 million dollars
92.3	(a)	(i)	$a = 5.98$, $b = 21.6$
		(ii)	a represents the average increase of number of visitors when the maximum temperature is increased by 1 degree Celsius b represents the expected number of visitors when the maximum temperature is zero degree Celsius
	(b)		45.5
92.4	(a)	(i)	$a = 6.85$, $b = 24.3$
		(ii)	a represents the average increase of the hardness of a metal ingot when its breaking strength is increased by 1 tonne per cm b represents the hardness of a metal ingot when its breaking strength is zero tonne per cm
	(b)		65.4

Exercise 93

93.1	(a)	(i)	5.5
		(ii)	4.5
	(b)		$r_s = 0.0588$
	(c)		There is a weak agreement between Ravi and Yannick

Your Practice Set – Applications and Interpretation for IBDP Mathematics

93.2 (a) $r_s = 0.867$
(b) There is a strong agreement between the two judges

93.3 (a) $b = 2$, $d = 3$, $f = 5$ and $g = 4$
(b) $r_s = 0.905$
(c) It is more likely for a team to be at a higher position in the league if it scores more goals

93.4 (a) $r_s = 0.0857$
(b) (i) $a = 1$, $e = 5$
(ii) $R_s = 0.9$
(iii) The value of the Spearman's rank correlation coefficient increases

Exercise 94

94.1 (a) The values of the first variable are not quantifiable data
(b) $r_s = 0.8$
(c) It is more likely for a type of ice cream to have a higher ranking if its selling price is higher

94.2 (a) The values of the data are not linear
(b) $r_s = -0.833$
(c) It is more likely to have a lower score in Geography test if the score in Chemistry test is higher

94.3 (a) The positions of teams in the league are not quantifiable data
(b) $p = 5.5$, $q = 7$
(c) $r_s = 0.946$
(d) 1, 2, 3, 4 and 5

94.4 (a) $r = 0.814$
(b) (i) $b = a - 2$
(ii) $g = a + 3$
(c) $r_s = 0.893$
(d) The rank of the score of the diver B awarded by the trainee judge is unchanged

Exercise 95

95.1 (a) (i) 0.860
(ii) $a = 0.00360$, $b = -0.626$
(b) 1.9 kg
(c) 2.41 kg
(d) 2019

95.2 (a) (i) 0.982
(ii) $a = 2.5625$, $b = 6.375$
(b) 68 kg
(c) 91.0 kg
(d) February 2019

95.3 (a) (i) 0.982
(ii) $a = 14.1$, $b = 188$
(b) 343
(c) 28.1
(d) 1984

95.4 (a) (i) −0.926
(ii) $a = -1.17$, $b = 58.8$
(b) 45
(c) 0.0866
(d) After 7.29 minutes

Chapter 23

Exercise 96

96.1
(a) H_0: The number of heads follows the assigned distribution
(b) 10
(c) 3
(d) $\chi^2_{calc} = 6.93$
(e) The null hypothesis is not rejected as $\chi^2_{calc} < 7.815$

96.2
(a) H_0: The last digits are evenly distributed
(b) 10
(c) 9
(d) 15.8
(e) The null hypothesis is not rejected as $\chi^2_{calc} < 16.919$

96.3
(a) H_0: The number of emails received on each day follows the assigned distribution
(b) $a = 4$, $b = 5$
(c) 5
(d) p-value $= 0.0907$
(e) The null hypothesis is not rejected as p-value > 0.05

96.4
(a) H_0: The outcomes follows the assigned distribution
(b) 30
(c) 6
(d) p-value $= 0.0000496$
(e) The null hypothesis is rejected as p-value < 0.01

Exercise 97

97.1
(a) (i) H_0: The examination results and the corresponding sections are independent
    (ii) H_1: The examination results and the corresponding sections are not independent
(b) 2
(c) $\chi^2_{calc} = 17.4$
(d) The null hypothesis is rejected as $\chi^2_{calc} > 5.991$

97.2
(a) (i) H_0: The nationality of children and the choices of their most favourite fruits are independent
    (ii) H_1: The nationality of children and the choices of their most favourite fruits are not independent
(b) 6
(c) p-value $= 4.57 \times 10^{-10}$
(d) The null hypothesis is rejected as p-value < 0.01

97.3
(a) H_0: The age of adults and their preferences are independent
(b) $x = 13$, $y = 39$
(c) 4
(d) p-value $= 4.70 \times 10^{-7}$
(e) The null hypothesis is rejected as p-value < 0.05

Your Practice Set – Applications and Interpretation for IBDP Mathematics

97.4   (a) H_0: The age of staffs and their number of investment bank accounts are independent
        (b) $x = 5$, $y = 4$
        (c)   8
        (d) p-value $= 0.326$
        (e)   The null hypothesis is not rejected as p-value > 0.05

Exercise 98

98.1   (a)   The lengths of fishes are normally distributed
        (b)   (i) $H_0: \mu_1 = \mu_2$
             (ii) $H_1: \mu_1 \neq \mu_2$
        (c) p-value $= 0.0800$
        (d)   The null hypothesis is not rejected as p-value > 0.05

98.2   (a)   The volumes of bottles of milk are normally distributed
        (b)   (i) $H_0: \mu_D = \mu_L$
             (ii) $H_1: \mu_D > \mu_L$
        (c) p-value $= 0.0300$
        (d)   The null hypothesis is rejected as p-value < 0.05

98.3   (a) $H_1: \mu_1 < \mu_2$
        (b) p-value $= 0.0276$
        (c) -2.09
        (d)   The null hypothesis is rejected as p-value < 0.05

98.4   (a) $H_1: \mu_1 \neq \mu_2$
        (b) p-value $= 0.906$
        (c) -0.120
        (d)   The null hypothesis is not rejected as p-value > 0.1

Exercise 99

99.1   (a)   (i) H_0: The punctuality of buses and the locations of bus stops are independent
             (ii) H_1: The punctuality of buses and the locations of bus stops are not independent
        (b)   3
        (c) $\chi^2_{calc} = 6.41$
        (d)   The null hypothesis is rejected as $\chi^2_{calc} > 6.251$
        (e)   (i) $\dfrac{1}{5}$
             (ii) $\dfrac{3}{20}$
             (iii) $\dfrac{19}{85}$
        (f) $\dfrac{119}{165}$

99.2   (a)   (i) H_0: The number of free lunches offered and the positions of staffs are independent
             (ii) H_1: The number of free lunches offered and the positions of staffs are not independent
        (b)   9
        (c) p-value $= 5.95 \times 10^{-7}$
        (d)   46.0
        (e)   The null hypothesis is rejected as p-value < 0.05

- (f) (i) $\dfrac{1}{25}$
  - (ii) $\dfrac{8}{25}$
  - (iii) $\dfrac{1}{8}$
- (g) $\dfrac{1139}{2475}$

99.3
- (a) (i) 8
  - (ii) 5.5
  - (iii) $1 \leq X \leq 5$
- (b) (i) $a = 100$, $b = 60$
  - (ii) 6.5
  - (iii) 3.91
- (c) $\dfrac{7}{50}$
- (d) (i) H_0: The number of dolls owned by a female student and her nationality are independent
  - (ii) H_1: The number of dolls owned by a female student and her nationality are not independent
- (e) 4
- (f) p-value $= 0.0623$
- (g) 8.95
- (h) The null hypothesis is not rejected as p-value > 0.05

99.4
- (a) (i) 25.5
  - (ii) 10.5
  - (iii) $1 \leq X \leq 10$
- (b) (i) $a = 30$, $b = 146$, $n = 500$
  - (ii) 14.42
  - (iii) 8.25
- (c) $\dfrac{14}{15}$
- (d) (i) H_0: The age of an interviewee and the number of yogurt parfait cups ate are independent
  - (ii) H_1: The age of an interviewee and the number of yogurt parfait cups ate are not independent
- (e) 6
- (f) p-value $= 1.44 \times 10^{-31}$
- (g) 158
- (h) The null hypothesis is rejected as p-value < 0.01

Exercise 100

100.1
- (a) (i) $H_0: \mu_1 = \mu_2$
  - (ii) $H_1: \mu_1 < \mu_2$
- (b) p-value $= 0.365$
- (c) -0.350
- (d) The null hypothesis is not rejected as p-value > 0.1
- (e) $\dfrac{3}{8}$
- (f) (i) $\dfrac{1}{3}$
  - (ii) $\dfrac{2}{3}$

100.2
- (a) The number of times for a person to smoke in a day are normally distributed
- (b) $H_1: \mu_1 > \mu_2$
- (c) p-value $= 0.00556$
- (d) 3.11
- (e) The null hypothesis is rejected as p-value < 0.01

- (f) (i) $\dfrac{1}{4}$
    - (ii) $\dfrac{1}{3}$
- (g) $\dfrac{7}{44}$

100.3
- (a) The scores in the assessments are normally distributed
- (b) $H_1: \mu_{A1} \neq \mu_{B1}$
- (c) p-value $= 0.0465$
- (d) 2.20
- (e) The null hypothesis is rejected as p-value < 0.05
- (f) (i) $\mu_{A1} < \mu_{A2}$
    - (ii) As p-value > 0.05, the null hypothesis is not rejected
- (g) $\dfrac{10}{21}$

100.4
- (a) The ball speeds of free kicks are normally distributed
- (b) $H_1: \mu_{R1} < \mu_{R2}$
- (c) p-value $= 0.0486$
- (d) -1.77
- (e) The null hypothesis is rejected as p-value < 0.1
- (f) (i) $\mu_{R2} \neq \mu_{M2}$
    - (ii) As p-value > 0.1, the null hypothesis is not rejected
- (g) $\dfrac{1}{6}$

CPSIA information can be obtained
at www.ICGtesting.com
Printed in the USA
BVHW011334170522
637236BV00006B/308